Populations of Individuals

Populations of Individuals

Understanding Biological Variation from Molecules to Ecosystems

Kent J. Bradford

Distinguished Professor Emeritus, Department of Plant Sciences, University of California, Davis

OXFORD
UNIVERSITY PRESS

OXFORD
UNIVERSITY PRESS

Oxford University Press is a department of the University of Oxford. It furthers the University's objective of excellence in research, scholarship, and education by publishing worldwide. Oxford is a registered trade mark of Oxford University Press in the UK and in certain other countries.

Published in the United States of America by Oxford University Press
198 Madison Avenue, New York, NY 10016, United States of America.

CIP data is on file at the Library of Congress.

ISBN 9780197846001

DOI: 10.1093/9780197846032.001.0001

Printed by Integrated Books International, United States of America

The manufacturer's authorized representative in the EU for product safety is Oxford University Press España S.A. of Parque Empresarial San Fernando de Henares, Avenida de Castilla, 2 – 28830 Madrid (www.oup.es/en or product.safety@oup.com). OUP España S.A. also acts as importer into Spain of products made by the manufacturer.

MIX
Paper
FSC FSC® C183721

Foreword

The dichotomy of populations and individuals spans every corner of the biological world. For the epidemiologist, thousands of individual patients make up the city experiencing a pandemic. For the physiologist, millions of individual cells comprise the tissue and determine its health or pathology. For the biochemist, billions of individual protein molecules are together the enzyme solution that transforms one substrate into another. While the relationships between these parts and the whole can seem simple, in reality they are often surprisingly subtle. Navigating these connections is both critical and an often overlooked area of need in the life sciences.

One persistent challenge is that academic disciplines have traditionally positioned themselves at the extremes of these relationships, with limited connections between the experts at the opposite poles. It is not uncommon for a cell biologist to have a deep knowledge of an individual cell's internal world, but to rarely look at a field of the same cells on a histochemical tissue slide. The physiologist who prepared the slide and who has built an intuitive grasp of the subtle patterns seen at the multicellular population scale, is most likely in a different department on the other side of campus, or on a different campus altogether. Similarly, a physician's daily view is filled with the individual patient in front of them, and while statistics may guide their overall approach, they are only a backdrop for many factors in that patient's history that make their clinical presentation unique. The epidemiologist who is keenly aware of a recent trend in patient histories for a given disease and who might have a growing interest in that particular combination of symptoms is likely at least a building or two away from the examining room. There are of course some very notable exceptions to this divorced state, not least of all because brilliant scientists who master both ends of a spectrum often become the leaders in their professions. For a successful few, the ability to transition from the individual to the population view and vice versa is almost a scientific superpower. Yet, these famous exceptions effectively prove the rule: it is much more common to be trained, and to remain for one's whole career, at one landing in the hierarchy of biological organization, without traveling far either up the staircase or down it.

Yet, another tradition runs counter to the divisions inherent in our academic approach to the different levels of life science. Some life sciences departments strongly encourage trainees to diversify their diet of seminar attendance and paper reading. The highest regard is given to cases where breakthroughs arrived only because an individual who had read widely and thoughtfully was able to connect two seemingly disparate systems. Yet, even here, the expectation to spend some time thinking about a different disease or organism is often informally communicated and is accompanied by the distinct disadvantage of, at least in the short term, distracting from one's immediate project. The present volume firmly embraces the tradition of hunting for common threads buried in the tangle of other fields. Here, Dr. Bradford draws on fields from the subcellular to the ecological, and from the very origins of life to the plant and animal kingdoms, to underscore and develop a wide-ranging thesis that oversteps traditional disciplinary boundaries.

The truly cross-disciplinary and integrative nature of the following work embodies the spirit of many efforts in recent decades to break down traditional barriers between fields. The vision of bringing experts at different levels of organization—physicians and epidemiologists, biophysicists and cell biologists, and so on—under the same roof has driven many new institutes and centers, often with some variation on "integration" in their name or mission statement. This blending of disciplines is often intentional, rooted in an intuition that a collision between two orthogonally situated worlds will inevitably impose areas of constructive and destructive interference and lead fortuitously to conceptual breakthroughs. Other times it is somewhat less guided, resulting simply from the semi-random physical juxtaposition of many disciplines on a large university or research campus. In either case, the discussion in this book is a reminder that such combinations have been, and continue to be, crucial engines of scientific progress. It will be important to keep such examples in mind even as political changes drive what may be the largest shift in higher education and research funding in nearly a century. As we enter this new era, losing sight of the benefits of cross-disciplinary fertilization poses a significant danger to innovation in the decades to come. Centers of broad multidisciplinary study, and the freedom of investigators to navigate at will across an open landscape of biological scales, will not be replaceable by siloed and hyperfocused private endeavors. It is my hope that this book, and others in this vein, will remind us that the often tangled complexity of the university, while sometimes frustrating, can be an enabling factor in the transformation of human knowledge. These complex intellectual ecosystems are not perfect, but they require great energy to build and will not easily be re-created once they are disassembled.

However, the potential impact of this work goes beyond simply connecting different disciplines; it envisions an example of the broadly useful concepts that we hope will emerge from our biology-wide interconnections. Science often needs to proceed by intuition before enough raw evidence has been gathered to form a proper hypothesis. If our universities and institutes are our experiments in the impact of organizational behavior on human knowledge, then the resulting literature must be considered as our results. The many thousands of individual papers resulting from our experiments in reorganization are our raw data. So far, there are countless anecdotes of fruitful new collaborations, a tantalizing payoff of the exploratory phase that is often the most exciting part of the scientific endeavor. However, despite these successes, it still feels that what is lacking is a coalescence of this raw data into an overarching set of concepts that guide our approach to connecting different levels of biological hierarchies. In the present work, we begin to glimpse the higher-order concepts that will emerge from these raw data, as Dr. Bradford stitches together unifying ideas in trans-scale relationships.

The timing of this work could not be more opportune, as simultaneous revolutions in data collection in multiple fields make it possible to view individuals and populations simultaneously. At the molecular level, many millions of molecules can now be collected in a single dataset by electron microscopy or tomography, powering the assembly of detailed structures from population or subpopulation statistics. In physiology, single-cell genomic or transcriptomic sequencing methods make it possible to catalog the thousands of individual cell states present within a tissue. The population of an entire nation can be filtered and subdivided by individual medical records and even hourly locations. We have gone from being ankle-deep in data to being well over our heads in little more than a decade. Navigating this data-rich world will require us to go far beyond the anecdotal thinking we are trained in and to which we have become well accustomed. The journey in the following pages offers a view of the intellectual vessels we will need as the tide comes in.

John G. Albeck

Preface

In introducing this book, I feel that I must share the comment I received from Bill Finch-Savage, a friend and colleague who kindly agreed to have a look at an early version of the manuscript: "Wow, I can see why you wanted to get this off your chest before fully retiring." So true. The initial insight came to me at the urging of Anna Tarquis, a PhD student in my laboratory in 1988. Anna was a far better mathematician than I am (and remains so as a Professor of Mathematics at Universidad Politécnica de Madrid), and she urged me to read a recent paper by Robert Gummerson analyzing seed germination using his new population-based model (Gummerson, 1986). I kept putting it off, but eventually her persistence got me to read the paper. As a seed biologist, I was entranced by the concept proposed there that individual seeds differed in their sensitivity to water stress during germination. I vividly recall standing in my office in front of a blackboard with the equations on it trying to understand how the model proposed by Gummerson could be utilized in our own work on seed–water relations. Suddenly the power of this idea hit me, as it could be applied to many aspects of seed biology, where diversity of behavior is a hallmark associated with dormancy mechanisms and environmental sensitivities. I subsequently focused on formalizing and exploring the consequences of this concept, which resulted in my first paper on the population-based threshold (PBT) model (Bradford, 1990).

Since that original insight, I have come to a view of biological data that is rather different from that of most biologists. In exploring the PBT model and discovering how well it describes seed behavior across a range of situations, I became convinced that its applications were not limited to seeds. As my eyes became entrained to the patterns generated by the PBT model, I began to see them occurring in data everywhere I looked. I actually do see biological data differently than most people, as I can recognize the underlying individual variation that may explain the observed responses. Over the past thirty-five years, I have collected examples and thought about how I might make my vision accessible to others. In the end, as Bill Finch-Savage noted, I was able to "get this off my chest" by writing this book after my retirement from the University of California, Davis. In this book, I take an evolutionary approach to explore how the underlying and largely uncontrollable (and nongenetic) variation present in biology could be adapted to and actually "harnessed" into what is a simpler and more automatic regulatory system than the one that is generally assumed to be present. By providing examples across the breadth of biology, I am attempting to give readers that same "Ah ha" moment that I had when it occurred to me that some of the fundamental principles that I had been taught as an undergraduate studying biochemistry might need some modification. I incorporate that revelation in a metaphor of "population-colored glasses" through which data can be viewed as having an underlying population basis. Simply changing the assumption that individuals at a given level of complexity (i.e., molecules, genes, cells, tissues, organs, organisms) are identical to the possibility that they vary within defined functional subpopulations associated with their sensitivity to perform their assigned biological roles radically alters what can be seen in the data.

It has been a rather daunting task to try to delve deeply enough into the broad scope of biology to provide believable current examples of and implications for adopting this population-based viewpoint. I have done my best, and no doubt experts in these fields will find some flaws in my interpretations. I can only hope that this will spur them to bring their own expertise to bear on further refinement, rather than abandoning the effort and failing to even put on the population-colored glasses. On the other hand, it is perhaps for the best that I was unable to take this effort on until retirement, which occurred shortly after single-cell, single-nucleus and single-molecule technologies were developed. These approaches can actually test the PBT approach by measuring individuals and quantifying their variation, which I argue is an alternative to the common "noise-based" explanations. Certainly, stochastic noise is present in biology, as it is in physics, but I cannot believe that biology developed to its current extent without devising mechanisms to contain and even utilize such variation to its advantage. Rather, I believe that in addition to genetic variation, molecular, cellular, physiological and developmental variation are features, not bugs, of biology. To present and support that contention is the purpose of this book, and I welcome readers to engage it with an open mind and a willingness to see differently what our modern technologies are revealing to us about the roles of individuals at all levels of biological organization.

I also need to thank many people who conducted research in my laboratory and have supported this effort in various ways. In particular, Peetambar Dahal and Pedro Bello were staff scientists in my laboratory who contributed crucial research to support and extend the PBT model in relation to seed germination. Many other graduate students and postdoctoral scientists also refined different aspects, including Veria Alvarado, Dominique Ardura, Louis

Boddy, Zhiyuan Cheng, Sunil Dutta, Alfred Huo, Shuangshuang Liu, Bing-Ri Ni, Hiro Nonogaki, Reagan Reed, Andres Schwember, Oscar Somasco, David Still and Anna Tarquis. I also am grateful to colleagues who have been both inspirations and contributors to the ideas in this book. I am particularly indebted to Anthony Trewavas, who responded positively to an inquiry from me in 1993 and remained a strong supporter throughout the development of this book. And as noted above, Bill Finch-Savage was an early adopter and promoter of the PBT concept in seed biology. I appreciated the support of my UC Davis colleague Judy Callis, who was not a seed biologist, but heard me describe the PBT model in a seminar and remained a solid supporter thereafter, including reading and commenting on an early version of the manuscript. I also greatly appreciate that another UC Davis colleague, John Albeck, shared unpublished data with me and agreed to write the Foreword from his perspective as a mammalian cell biologist. Many other colleagues around the world also deserve my thanks for their collaborations and contributions, including Phil Allen, Esmaeil Bakhshandeh, George Bassel, Diego Batlla, Roberto Benech-Arnold, Derek Bewley, Eduardo Blumwald, Benjamin Bradford, Julia Buitink, Liana Burghardt, Daniel Côme, Marc Cohn, Françoise Corbineau, Kathleen Donohue, Albert Fischer, Maraeva Gianella, Filippo Guzzon, Tony Haigh, Fiona Hay, David Horvath, Zhenying Huang, Imtiyaz Khanday, Francois Korn, Olivier Leprince, Johan Leveau, Maeli Melotto, Susan Meyer, Hugh Price, Daniel Runcie, Santosh Sanjel, Johanna Schmitt, Breanne Sparta, Larry Venable and many others. The following friends and colleagues also provided feedback for versions of this book from their diverse viewpoints: Dominique Ardura, Roberto Benech-Arnold, Ed Clemens, Carroll Cross, Henk Hilhorst, Ger Lenssen, Adrienne Roeder, Howard Shapiro, Sharon Welch and David Zilberman, for which I thank them for their time and interest. The assistance of editors Katie Helke and Jessica Neuberg and production manager Hemasankar Arumugam at Oxford University Press in bringing the book to publication is gratefully acknowledged. I also thank Robert Saigh for preparing the index. I appreciate the love and support of my wife Barbara Zadra and my sons Ben and Jesse who have made my life wonderful. I apologize to anyone I may have inadvertently missed, and of course, any errors remaining in the book are my responsibility.

Kent J. Bradford

Contents

"A theory is the more impressive the greater the simplicity of its premises is, the more different are the kinds of things it relates, and the more extended is its range of applicability."

—Albert Einstein

Autobiographical Notes (1946), p. 33. Quoted in Elkana, Y. (1982) The myth of simplicity. pp. 205–251 in Holton, G. and Elkana, Y. (Eds) Albert Einstein, Historical and Cultural Perspectives: The Centennial Symposium in Jerusalem. Princeton University Press, Princeton, New Jersey, United States.

1
Populations of Individuals

Variation is a fundamental feature of biology, whether at the molecular, genetic, physiological, cellular, organismal or ecological level of organization. This underlying variation has long been accepted as the basis for natural selection and evolution, without which "nothing in biology makes sense" (Dobzhansky, 1973, p. 125). The importance of genetic variation was established by Mendel and remains the basis of modern genetics and breeding. However, in other areas of biology, such as physiological or cellular biology in which genetic variation within a given individual organism is minimal, variation is more often considered to represent either sampling and measurement error or random biological "noise" (Raser and O'Shea, 2005). Even as methods have been developed to investigate biology at deeper and deeper layers, such as by sequencing the mRNA molecules (transcriptomes) of individual cells (Karaiskos et al., 2017), terms such as "transcriptional noise" have become common in light of the wide range of variation observed at those levels (Blake et al., 2006; Ham et al., 2021). However, how could it be that a phenomenon as complex and successful as all of biology could be based on an inherently "noisy" process at such a fundamental level? Long before Mendel, people recognized that children resembled their parents, although they had no idea of the genetic mechanisms of trait transmission. While there is still much to be learned about the details, it is clear now that the transmission of traits encoded in DNA occurs in a way that largely explains such observations. Similarly, what is currently called biological "noise" might also have a relatively straightforward explanation that both clarifies its basis and explains its apparent persistence.

We must therefore examine the nature of biological variation in general, particularly within genetically homozygous individual organisms where genetic variation is not the answer. Is biology simply a loosely organized system that tolerates a remarkable amount of uncontrolled random (i.e., "noisy") variation at all levels, and yet is somehow consistently able to reproduce cells and organisms that constitute all of the distinct species that have lived over the past almost four billion years? Or is it just the crudeness of our measurement tools, resulting in sampling and experimental error that requires statistical

Populations of Individuals. Kent J. Bradford, Oxford University Press. © Kent J. Bradford (2026).
DOI: 10.1093/9780197846032.003.0001

tools to mine out the “true” average behavior and ignore the apparent noise? Or could it be something else, a system in which ubiquitous random variation has been incorporated as a fundamental and integral part of the functional machinery of biology?

The purpose of this book is to explore the latter thesis that variation among individuals, possibly random or stochastic at its base, is inherent in all levels of biological complexity and function and has, in fact, become a fundamental mechanism underlying how biology works, providing an explanatory framework for key features of biological systems across scales. Models that describe such variation can be used to both quantify and predict the behavior of biological populations and to transfer information between different levels of biological organization. The reader may ask, could such a broad theory really underlie our understanding of biology, hidden in plain sight under our microscopes and in our laboratory and field observations? I am risking hubris to even propose such a possibility. Yet, nothing in science is evident until it is first conceived, proposed and then tested through observation and experiment. Thus, it will be the goal of this book to propose this thesis, describe its properties, provide a range of examples across the scope of biology and discuss some implications if it is, in fact, true. Of course, whether it is true will be based upon further targeted experimentation to test its validity across the scope of biology, hopefully stimulated by this book.

Engaging in this process may require a temporary suspension of belief by the reader regarding what they may already feel is true and an openness to seeing biological processes in a different way. This can be considered as providing a new Kuhnian paradigm (Kuhn, 1962), or perhaps just curiously “population-colored” glasses, through which to view biological data and its interpretation. While reading this book, imagine that you can put on glasses that enable you to “see” populations of diverse individuals existing and underlying the observed features or behaviors. These populations will generally be organized into the familiar bell curve (or Gaussian distribution). As noted by Pinker (2021, p. 205), “Bell curves . . . are common in the world; they arise whenever a measurement is the sum of a large number of small causes, like many genes together with many environmental influences.” Such distributions are so common that they are called “normal” distributions, and in fact, most statistical tests assume that the observed data match such a distribution or require the transformation of the data into such a form in order for the analyses to be valid. Thus, our faith in statistical analysis itself is based on the existence of such underlying population distributions. It is evident from Pinker’s definition that biology provides just the conditions that will result in normal distributions, that is, thousands of genes in trillions of cells interacting with a wide diversity of molecular, physiological and environmental inputs. However, the statistical analyses of such distributions in actual data have devolved into questions such as whether the means of two such distributions can be distinguished (regardless of their extent of overlap) or the probability that one such distribution differs from an imagined null hypothesis (Clayton, 2021). We have somehow been able to ignore that the distribution represents the actual frequencies of individual molecular, cellular and organismal entities across the spectrum of observed variation. Our statistical interpretations may be convenient and useful, but they are secondary to this underlying fact. As such, I will dwell little here on statistics per se and focus attention on the distributions themselves, which represent the actual biological variation for the traits or processes of interest.

Before continuing further, I must acknowledge that even the concept of "individual" can be problematic in the practice and philosophy of biology, given the interlocking layers of systems that must function for life to continue (Trewavas, 1999; Bueno et al., 2018). It is evident that since the initiation of self-replicating biological entities, multiple players have been shown to be involved, from nucleic acids and protein molecules to cells, tissues, organs, organisms, species and communities. As such, it can be difficult to identify where the "individual" is in this network. At the extreme, Hermida (2016) has argued that all of life on earth is only one "life-individual," as apparently all of existing life is connected back to a single initial origin and we have no other contrasting examples of how life might possibly work. On the other hand, Wong et al. (2023, p. 7) recently proposed a new fundamental "law of increasing functional information" in which the "functional information of a system will increase (i.e., the system will evolve) if many different configurations of the system undergo selection for one or more functions." The "many different configurations of the system" would seem to be consistent with what we would term "individuality" or at least variation among individuals at each level of complexity under consideration. These authors propose that the functional interactions of such individual system configurations constitute a new fundamental law that is on a par with those of thermodynamics. However, while both individuality and variation are ubiquitous in biology, how these components interact functionally remains largely unexplored. At the molecular level, physical processes such as diffusion and viscosity play a major role, while at higher levels of organization (and larger physical distances) other biological processes become primary (Parthasarathy, 2022). What I will be proposing here is that the key to understanding how this works is to view individuals as constituents of related populations in which the pattern of functional or behavioral variation among those individuals is a fundamental feature of the system. Based on the discussion above, I will in general be assuming that such variation is organized into normal distributions, although this is more for convenience than a requirement.

In addition to highlighting populations of individuals, I will also be focusing particularly on the response thresholds for specific behaviors that are represented among such individuals. That is, individuals can differ in their sensitivities to respond to given factors, whether they are chemicals, hormones, environmental conditions, other organisms and so on (Williams, 1956; Bradford and Trewavas, 1994; Trewavas, 1999, 2012; Bradford, 2018). As such, the models that I will be describing are called population-based threshold or PBT models (Bradford et al., 1993). A key distinguishing feature of individuals in such a population is that their thresholds to respond to a factor of interest differ. Once the factor level exceeds the response threshold of a given individual, that individual will perform the function programmed to be elicited by the factor. In many cases, the individual response may be quantal, or on/off, such as the electrical firing of a neuron or the germination of a seed. As the factor dosage increases, more individuals in the population will respond as their thresholds are exceeded. The overall response will appear as a sigmoid increase due to the recruitment of additional individuals whose thresholds vary in a normal distribution. In addition, as the factor level increases above the threshold of a given individual, the time for it to respond is shortened proportionately. Thus, in addition to making use of the inherent variation present, which was the source of concerns about "noise," the model also accounts for the shape of most dose-response curves and for the kinetics of processes such as enzyme activities across a wide range of substrate concentrations. I will provide examples of such behavior across all scales of biology. If you doubt that such a simple threshold-based mechanism could possibly underlie the complexity of all of biology, see Braitenberg (1984). There you will find an entertaining exploration of how the basic requirements and concepts of neuronal function, the most complex biological phenomena known (Damasio, 1999), can be mechanistically described simply on the basis of individual receptors exhibiting a range of threshold sensitivities and their assembly into interaction networks that determine distinct outcomes.

In the following chapters, I will describe the concepts underlying PBT models and how they can explain behavior in response to relevant molecular, metabolic, physiological and developmental factors. The remarkable behavior of plant seeds will figure prominently, as this is the biological system in which the PBT approach was conceived and has been explored most fully (Gummerson, 1986; Bradford, 1990; Bradford and Bello, 2022). Seeds are positioned conveniently in the biological hierarchy, as they inherently consist of populations of individuals. It is the biological function of seeds to disperse individually and independently to propagate the genes (and cells) of their parents into a new generation of plants. There can be little philosophical objection to considering them as individual biological organisms. At the same time, they can exhibit a bewildering array of expectations, anticipations, requirements and responses with respect to their interactions with the environments in which they find themselves (Baskin and Baskin, 2014, 2021). These interactions have been honed through selection to enable germination of an individual seed to occur when environmental conditions are more favorable for survival of the resulting seedling, thereby increasing the likelihood of propagating itself and the species of which it is a representative. Those diverse germination responses, in turn, are controlled by hormonal signals that are sensitive to specific environmental inputs (e.g., temperature, water, light, nutrients). This provides a system in which response distributions to a wide array of physiological and environmental factors can be tested by observing the relatively simple phenotype of the germination times of individual seeds in specific populations (seed lots). From this PBT vantage point, which is clearly based on the collective behaviors of individuals, the behaviors of molecules and cells in microorganisms, plants and animals can be examined. Beyond the individual organismal level, ecological and evolutionary concepts based on species as the common genetic unit also fit well with PBT approaches (Trewavas, 2014; Donohue et al., 2015; Gianella et al., 2021). Even societal levels of organization, such as economics and politics, can be viewed through a population-based lens (Zilberman et al., 2012).

The need for such a conceptual or organizational landscape is evident in the revolutionary impacts that single-cell and spatial transcriptomics, proteomics and metabolomics are having on our understanding of biological complexity (Bressan et al., 2023, p. 11):

> Just a few years ago, the ability to map the molecular make-up of every cell in a tissue, and interrogate hundreds of markers and millions of cells in the same experiment, would have been considered more science fiction than science. Now, single-cell and spatial omics technologies are advancing at breakneck speed and are granting scientists new powers to investigate complex biological processes. With this power comes the responsibility of assessing the limitations of this new breed of methods, their ideal application niches, and the best way to apply good experimental practices and rigorous analysis to the data that is produced. As more and more laboratories get access to these new-generation techniques, there is no doubt that incredible biological insights are just around the corner.

While such individual results initially seemed disconcertingly "noisy" (Stadler and Eisen, 2017), it has become clear now that this was because earlier methods had pooled together large numbers of distinct cell types in tissue-scale samples and accepted mean values as the underlying reality. Now, however, single-cell and spatial omics methods have revealed at least 3000 distinct cell types in the human brain, based upon characteristic variation in their transcriptomes and associated with their spatial locations and their functional diversity (Johansen et al., 2023; Siletti et al., 2023); similar results are becoming available across molecules, microbes, tissues, organs and species (Kanatani et al., 2024; Rood et al., 2024; Kapanidis et al., 2026). While it will neither answer all questions nor satisfy all criticisms, viewing such systems as being organized into populations of related but systematically varying individuals provides an alternative way to extract consistent and meaningful biological functionality from such diversity.

The final chapter of the book will examine some of the implications if biology is based upon and constructed around individual variation organized into populations rather than around means or averages and their associated "error" or "noise." The primary implication is the recognition that variation is a feature, not a bug, in the hardware and software of biology. It is neither error nor noise, but rather a highly sophisticated mechanism for regulating interactions in an extremely complex system composed inherently of collections of diverse individuals. Think of jazz music: it is based on individual improvisation on a common theme, which is also a fundamental feature of biology. The PBT concept endows the system with plasticity, flexibility and adaptability, but also stability, regulatory simplicity and functional integration in response to physiological and environmental signals across organizational levels. I will also highlight some major implications for experimental design and interpretation if this concept is true. Rather than measurements at the individual level revealing noise, it is the pooling of populations of individuals into bulked samples that provide only averaged values that is misleading (Williams, 1956). The current trend toward measuring individuals (at all levels of organization) and subsequently analyzing them as members of distinct subpopulations is the only way to avoid this obfuscation by pooling. We must therefore be consciously aware of this potential pooling error as we plan observations and experiments and collect, analyze and interpret such data. The goal of this book is to make a start toward achieving that longer-term goal.

As this book discusses data and examples from across the span of biology, it may be difficult for all readers to understand all parts with equal facility. Thus, I can suggest that another approach is to read Chapter 2, which describes population-based models in general, and then go to the chapter that is closest to your own expertise, whether that is molecular biology, microorganisms, plants, animals, developmental biology or ecology. Then, hopefully, your interest will be sufficiently piqued that you will go back and read the remaining chapters to get the full evolutionary sweep of the concept. Of course, you can also read the chapters in order and see how the concepts illustrated may apply to your own field when you get to it. Either way, just put on your population-colored glasses, and let's have a look at what we can see.

2

Population-based Threshold Models

Concepts and Features

Populations and distributions

It is now evident that biology is embodied in populations of individual cells (Mukherjee, 2022; Arias, 2023). Single-celled organisms are individuals, and a key story in evolutionary biology is how they were able to band together into the multicellular systems of tissues, organs and organisms that constitute the full biological universe (Pennisi, 2018). Variation among individuals is fundamental to genetics and natural selection; otherwise "survival of the fittest" would not have persistent effects in following generations. The random segregation of chromosomes in sexual reproduction results in normal (or Gaussian) distributions of phenotypic traits determined by the transmission of the genes and their DNA sequences on the chromosomes. How this occurs can be seen using a simple example of the outcomes of throwing a pair of dice multiple times (cf., Pinker, 2021). Each die has six sides, and if it is fair, the likelihood of each side showing up after a toss will be equal. However, if two such dice are thrown and their values are combined, the likelihood of different totals is based on the number of ways those values can be achieved. This is illustrated in Figure 2.1, which shows the possible ways that two dice can be combined and the sums of the numbers on each die. For example, there is only one way for the totals of 2 and 12, that is, two 1s or two 6s. In contrast, there are six ways that the two dice can total 7. Thus, in repeated tosses, a total of 7 will occur more frequently than any other total. In fact, the larger the number of tosses, the more closely the diagram in Figure 2.1 will resemble a normal distribution, with the most frequent totals in the middle and the least frequent ones at the extremes. In this case, assuming the dice are fair, there will be an equal number of tosses with totals above and below 7, which is therefore the median value (half higher and half lower). In addition, as the distribution is symmetrical in this case, 7 will also be the mean (average) value across all the throws.

Populations of Individuals. Kent J. Bradford, Oxford University Press.
DOI: 10.1093/9780197846032.003.0002

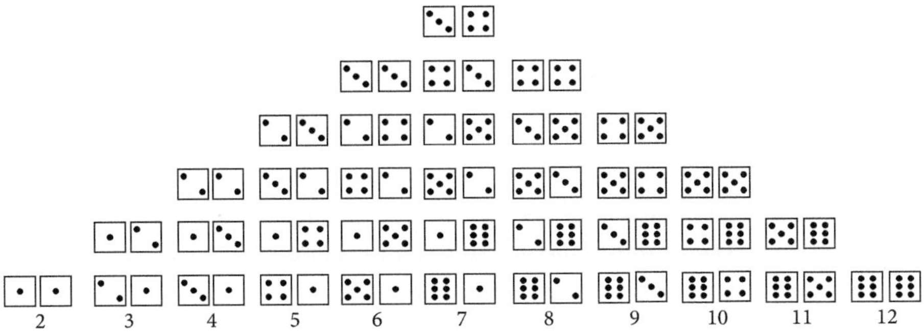

Figure 2.1 Illustration of the possible outcomes of two six-sided dice thrown multiple times in terms of the total value shown on both dice. There is only one way to total either 2 or 12, but multiple ways to total the numbers in between, with the most possibilities totaling 7. With only a few throws, any of these combinations could occur, but as the number of trials increases, the outcome will approach the normal distribution.

If these outcomes are so predictable, how can dice be used in gambling? In fact, much of statistics is based on attempting to predict the probability that a given result will occur on any single throw of the dice (Clayton, 2021). The outcome of any single toss is random, or stochastic, and it is essentially unpredictable in isolation, as any gambler knows. However, as the number of throws increases, the probability that the distribution illustrated will occur increases essentially to 1, or virtually a certainty. Thus, while the outcome of any single toss of the dice is unpredictable, the normal distribution of values around the mean is inevitable with enough tosses. It is important to note that no measurement error is involved here. There is no uncertainty about the actual outcome of an individual toss, even though it was unpredictable beforehand. Each of the combinations shown in Figure 2.1 is real, and it can and will occur at the expected frequency with enough trials. We would also be correct to say that with high numbers of tosses, the sum equal to 7 would appear with the highest frequency. However, this mean value alone does not fully characterize what has occurred during the trials.

Similarly, when chromosomes randomly segregate during meiosis, forming the sexual gametes, and the parental chromosomes in each gamete then rejoin in fertilization, there are many possibilities for which chromosomes will be present in the fertilized egg. This will depend upon the number of chromosomes characteristic of the species. For humans, with twenty-three pairs of chromosomes, the possible combinations are 2^{23} or 8,388,608. With regard to the dice example, this indicates that it is very unlikely for two siblings to inherit identical genomes (unless the zygote forms two embryos after fertilization, as in identical twins). In fact, when following a particular trait that differs between the parents, the inheritance pattern is quite simple if the gene responsible for the trait from one parent is dominant over the other. In the first generation, the dominant trait will appear in all the progeny, and if that individual can fertilize itself (as in plants), a 3:1 ratio of the trait will appear in its progeny (i.e., grandchildren of the first generation), referring to the dominant + mixed (heterozygous but expressing the dominant phenotype) and recessive phenotypes. In his successful experiments with peas, Mendel had focused on traits (e.g., wrinkled or smooth seeds) that were transmitted by this simple single-gene dominant vs. recessive system (rather than more complex polygenic traits). This is the minimal relationship that would result from flipping two coins and recording the results as both heads (1), both tails (1) or one of each (2). It is nice and simple for breeders when it works this way, but with more genes involved and more complex traits, the possible outcome categories and ratios become more complicated. It would therefore not have been possible for Mendel to formulate his genetic model from the data he would have been able to collect. In fact, most traits result from minor or overlapping contributions from multiple genes. In this case, the distribution of phenotypes for the trait looks more like Figure 2.1, and in fact it will approach a normal distribution as the number of contributing genes increases. In breeding, these genes of minor effect are called "quantitative trait loci" or QTLs, which can be detected by phenotyping large numbers of segregating progeny and identifying the regions on specific chromosomes that carry the causal gene(s). In QTL analyses, rather than an on/off response as in the dominant/recessive case, a genetic locus that contributes only 5% of the final trait effect can be identified as significant. Its effects on a phenotype can be captured and incorporated into future plant or animal varieties by repeatedly selecting the extreme fraction of individuals exhibiting the trait in a population to increase the frequency of this allele (gene variant) in the breeding population.

When the variation in a population is due to genetic diversity, as in progeny from a cross between genetically distinct parents, it is simple to see how this will result in trait variation in the progeny as those distinct genes segregate. However, this scenario will not be our primary focus here. Within a given organism, which arose either from division of a prior cell or from the combination of an egg and a sperm cell, all of its cells theoretically (i.e., excluding random mutations) carry an identical set of chromosomes and the genes they contain. A major question in biological development is therefore: how can the vast diversity of cell types, tissues, and organs in an individual organism be formed if they all have the same genome? To a large degree, this is done by opening (expressing) or closing (repressing) large fractions of the genome that are involved in specific developmental stages or physiological processes. In flowering plants, for example, more than 20% of the total genes in the genome may be expressed only during the reproductive stages of development, that is, flowering and seed/fruit formation. These organs are not present at other times during the plant life cycle (e.g., vegetative growth of roots and leaves), and therefore genes coding for their formation and function are not expressed until they are needed to form the reproductive organs. On the other hand, genes encoding proteins involved in more common processes, such as respiration, may be expressed in all cells and at all times. In animals, similar processes lead to the specialization of organs dedicated to specific tasks (e.g., heart, lung, liver, brain). How this selective regulation of gene activity across the same genome is controlled is a key question in developmental biology, and we will return to it repeatedly.

A useful characteristic of the normal distribution is that the fractions of the population represented by different sectors of the distribution are always constant relative to the width of the curve, which is determined by a factor called the "standard deviation" or sigma (σ) (Fig. 2.2). The standard deviation is a measure of the amount of variation in the values for the feature of interest among the individuals in the population; the larger its value, the greater the variation (the wider and shorter the distribution curve), and vice versa. However, regardless of the value of σ, the fractions of the population represented by standard deviation units away from the mean remain constant. Thus, if the mean and standard deviation are known, the shape of the corresponding normal distribution is fully defined. In Figure 2.2, the x-axis is labeled in units of σ, also known as "probits" (probability units) or "normal standard deviates." While recognizing that many other population distributions have been defined for specific situations, we will primarily focus on the normal distribution here.

Quantal (all or none) phenomena

At this point, I should also discuss "quantal" or "bimodal" phenomena, which are now often referred to as "digital" phenomena (Angel et al., 2015). Quantal events exhibit all-or-none, or on/off, behavior. A standard light switch is quantal or bimodal (on/off), whereas a dimmable light switch can take many different values between zero and the maximum, representing "analog" or "graded" behavior. Since data in computers are encoded as bits (either 1 or 0),

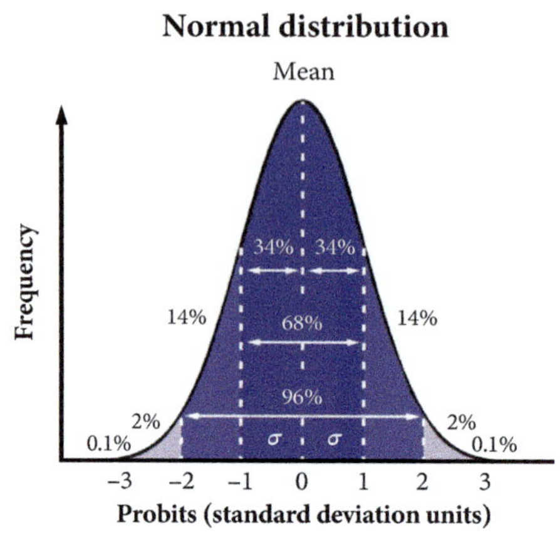

Figure 2.2 The normal (or Gaussian) distribution. It is defined by the mean (μ) and the standard deviation (σ) according to the equation:

$$y = \frac{1}{\sigma\sqrt{2\pi}} e^{-\frac{(x-\mu)^2}{2\sigma^2}}$$

The standard deviation is calculated by summing the squares of the differences of the trait values of each individual from the overall mean, dividing that sum by the number of individuals in the population, and then taking the square root, or

$$\sigma = \sqrt{\frac{\sum (x_i-\mu)^2}{N}}$$

where x_i is an individual value, μ is the mean value and N is the total number of individuals. The x-axis is labeled in standard deviation units, also known as probability units or "probits." One σ (or probit) on each side of the mean encompasses 34% of the population, or 68% within one σ on both sides of the mean. The next σ from the mean encompasses an additional 28% of the population, or 96% is within 2σ of the mean, and 99.7% of the population would be within 3σ of the mean. As the value of σ increases or decreases, the breadth of the curve will increase or decrease, but these fractions of the total population within standard deviation boundaries will remain the same.

Reproduced with permission from Bradford, K. J. and Bello, P. (2023). Understanding seed behavior: populations of individuals. *Acta Horticulturae* 1365: 1–16. https://doi.org/10.17660/ActaHortic.2023.1365.1.

their basic functioning is also quantal, leading to use of the "digital" term for on/off behavior. For example, seed germination is a quantal event in that at a given time after providing water (imbibition) an individual seed has either germinated or it has not. Other phases of plant growth, however, are more continuous, or analog, such as stem growth or root elongation. Animal physiology is generally considered under the analog category, although we will reexamine this later. For example, gene expression is often considered as an analog system, with expression of a given gene increasing or decreasing in proportion to the need for its product. However, the actual expression of a gene, that is, the synthesis of an mRNA molecule using the DNA sequence of the gene as a template, is itself a quantal event based upon an RNA polymerase complex traversing the coding region of the gene from one end to the other (see Fig. 3.1). The only way that the rate of transcription can be increased is to have more RNA polymerase molecules move along the gene sequence within a given time. Thus, the sum of underlying quantal events often appears to be analog over time (see Fig. 4.4). As another example, heat stress triggers the expression of protective heat-shock proteins, which accumulate during exposure to high temperature. It would be expected that a general environmental signal such as temperature would trigger all cells uniformly, with transcription rates of heat-shock proteins in each cell increasing with time as the stress increases. However, a recent study showed clearly that in fact, a given cell triggered by heat will express heat-shock proteins at a constant level, and that the increasing amounts of these proteins in a tissue over time is due to increasing numbers of cells being triggered to produce them (Alamos et al., 2021). Thus, at the individual cell level, the switch may be quantal, but at the tissue level it appears as a quantitative analog response due to the increasing numbers of cells being triggered in an on/off manner (Tsimring, 2014; Davies et al., 2020).

Population-based threshold models

Variation in response thresholds

The population-based threshold (PBT) model is based upon the concept of response thresholds. In general, a threshold indicates that the magnitude or intensity of some factor or requirement must exceed a specific level for a given reaction, phenomenon, condition or result to occur. We can express this as the difference between the factor level (X) and the response threshold or base value (X_b), or $X–X_b$. When this value is positive, that is, the factor level exceeds the basal response threshold, the response will be triggered. However, as will be illustrated with data subsequently, the response thresholds can differ among individuals, generally according to a normal distribution. Thus, to describe the population behavior, we will use the term $X_b(i)$ to indicate the distribution of base or threshold values for individuals or fractions i across the population. For example, if the inducing factor level (X) is equal to the mean of the sensitivity distribution ($X = X_b(50\%)$), half of the individuals will be triggered and half will not be. That is, $X–X_b(50\%)$ will be greater than zero for 50% of the individuals in the population but less than the triggering thresholds of individuals in the other half.

This simple relationship is the basis for one of the key features of the PBT model, which is the ability to recruit or sideline individuals from a potentially responding population depending upon the factor level. Assume that Figure 2.2 represents the distribution of response thresholds of individual cells to factor *X*. As the factor level increases from left to right, an increasing fraction of the total population will be triggered to respond as the level of factor *X* increases above their individual response thresholds. Alternatively, if the level of *X* is within the range of thresholds, lowering its level would potentially turn off responding individuals, assuming that this is possible (i.e., it is a reversible process such as a sensor signal). This provides a simple and automatic system for adjusting physiological responses to varying levels of a signaling factor. The response of each individual in the population could be quantal, either on or off, depending upon whether its response threshold is exceeded by the level of factor *X*, but due to the population variation in threshold levels, an apparently quantitative (analog) response to changes in *X* will be generated automatically (see Fig. 4.4). The shape of the threshold distribution will indicate the relationship of the response to the factor level (Fig. 2.3A). For example, a distribution with a large standard deviation would result in a gradual change in response over a wide dosage range, while a narrow distribution with a small standard deviation would give a sharp and higher magnitude response once the factor level entered its response range (Fig. 2.3B). One could imagine that the former would be useful for regulating processes that can tolerate a considerable range of variation (e.g., seed germination in example above), while the latter could be a system in which higher levels of the factor indicate a situation that must be addressed within a small tolerance range (e.g., heat-shock response to elevated temperature).

To borrow the classic understatement of Watson and Crick in their original paper describing the structure of DNA (Watson and Crick, 1953, p. 737), "it has not escaped our notice" that the sigmoid curves in Figure 2.3B are ubiquitous in virtually all types of biological data dealing with dose-response relationships. Thumb through any issue of a biological science journal and you will see curves of this type frequently. This sigmoid response curve is generally accepted without question as to its underlying basis, and it is more or less unconsciously attributed either to the random effects of substrate or signaling molecule concentrations on diffusion rates or other physical laws, or in other cases to sampling error or "noise." However, in biological systems such response curves are highly reproducible, can span logarithmic orders

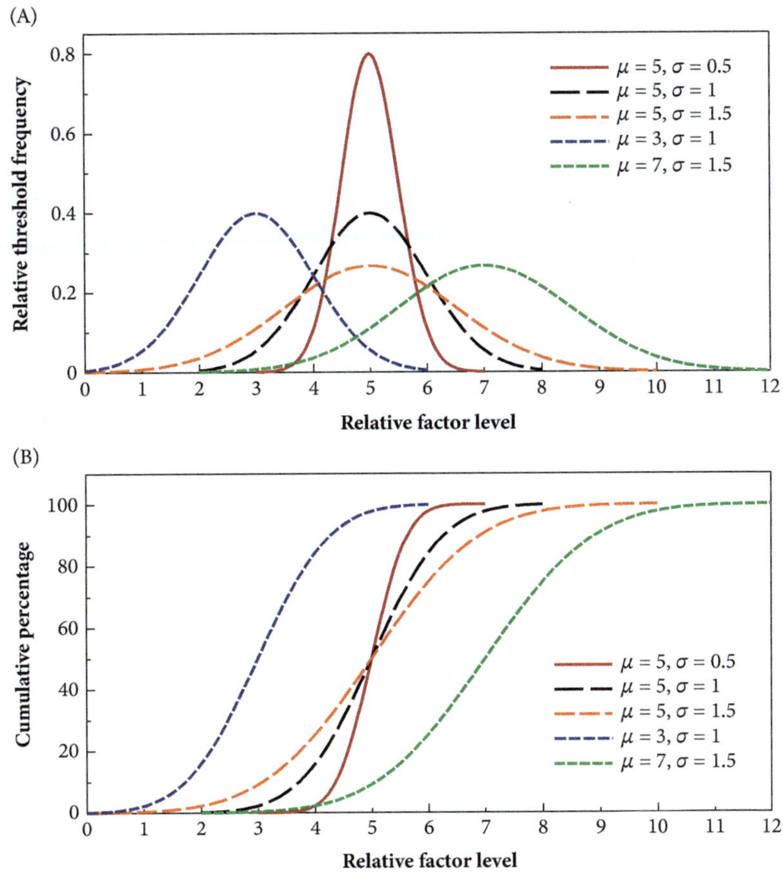

Figure 2.3 Threshold frequency distributions and cumulative response curves. **A.** Normal frequency distributions illustrating the effects of variation in the mean (μ) and standard deviation (σ) on their shapes. The frequency distributions are scaled by $1/\sigma$ so that the total areas under the curves are equal. As σ increases for a given μ, the sensitivity distribution broadens (from red to black to orange curves), while shifting μ moves the entire distribution to be more (from black to blue curves) or less (from black to green curves) sensitive to a given factor level, and changes in μ and σ can be independent. **B.** Cumulative curves corresponding to the distributions in **A** indicating the symmetric sigmoid shape of normal distributions when expressed as cumulative percentages of responders in the population as the factor level increases.

of factor magnitude, and are evident across organizational scales at which diffusion would not be explanatory, as will be shown later. The basic flaw in our thinking is to assume that the enzymes, receptors, cells, and so on, that are responding to the increasing substrate or signal factor levels are all identical in their sensitivities, so the variation observed must be due to something else.

What if, in fact, the responding entities vary in their inherent sensitivities and the sigmoid shape of the cumulative curve is a result of recruiting an increasing fraction of respondents from the pool exhibiting a normal distribution of sensitivity thresholds? Concentration and viscosity effects on diffusion could contribute to this, but what if the underlying basis is a direct result of sensitivity variation among the responding entities? Recruitment of additional responders across a normal distribution of response thresholds would automatically generate such curves, which can vary over a wide range by simply changing σ and μ (Fig. 2.3).

Diffusion and molecular kinetics

Before going further into the PBT model, some issues relating to diffusion and molecular kinetics should be discussed. Diffusion is a fundamental feature of the biophysical underpinning of biology (Parthasarathy, 2022). It is driven primarily by concentration differences of molecules across distances and the temperature and viscosity of the medium. As molecules move randomly, they will tend to move to where their concentration is lower, or down a concentration gradient. A simple demonstration of this feature is to use electrophoresis to move a dye band to the middle of a gel column, and then just turn off the power and let it diffuse. If everything diffused equally, you might expect that the borders would move outward in a line and the entire dye space would dilute. However, measuring the concentration of the dye over time results in a shape over space away from the original location that looks like a normal distribution, with fewer dye molecules at the extremes and increasing in density toward the middle of the band (origin). Those dye molecules that were nearest the original edge of the dye band would encounter the most extreme concentration gradient at the boundary, so their movement down that gradient would initially be rapid. They would also keep moving as more molecules moved in behind them, as the concentration will always be greater behind them than ahead of them. Those dye molecules near the middle of the dye band would experience no gradient initially and so would not exhibit net movement until the change in concentration had reached them. Near the middle, the gradient over distance would always be less, so directional movement would be less rapid, while on both edges of the band, the gradient would be maximal initially, and as it dissipated, those molecules that were in the lead would stay there. There would be no way for later-moving molecules to catch up with them, as the gradient would be continually decreasing everywhere away from the middle of the band. This results in exponential declines in concentration (and diffusion rates) away from the original center (mean) as the distribution spreads (standard deviation increases). Thus, diffusion has its effects largely at the molecular scale where the distances are very small (Parthasarathy, 2022).

Enzymes and molecular receptors function at the molecular scale at which diffusion is relevant. Substrates move into and out of enzymes' binding sites due to diffusion. Signaling molecules join with their receptors by diffusion. (However, cells have developed additional ways to move molecules and organelles across distances too large for diffusion alone to be effective; see Chapter 5.) It has generally been assumed that once individual enzyme or sensor proteins are made, they are essentially identical, so changes in enzyme or signaling turnover must be controlled by the substrate or signal concentration and diffusion (see Fig. 7.1). However, it is now clear that the structures of individual protein molecules in a cellular environment are not identical, despite having the same amino acid sequences, and more than 100 post-translational modifications to proteins have been documented that can affect their structures and activities (see Chapter 3). Thus, when enzymes or sensors in a location have different binding characteristics or response thresholds, those with the lowest thresholds will be the first to respond and might be quickly saturated, operating at their maximum turnover capacity. If the concentration is higher, there will be excess substrate/signal to trigger (recruit) additional enzymes/sensors having higher thresholds. So, if we could follow individual enzyme molecules, we would see that initially all the products would come from the most sensitive fraction of the population, but this represents only a small fraction of the total, like the dye molecules initially on the boundary as described above. As substrate/signal concentration increases, more enzymes/sensors with increasing thresholds will be recruited, which will act slower individually because the substrate concentration in excess of their own thresholds will be smaller.

The classic Michaelis–Menten kinetics used to characterize enzyme activity actually shows this pattern, with very rapid increases in turnover rates (steep slopes) at lower substrate concentrations that then decrease and asymptote at concentrations that saturate the system. However, in practice, the primary feature extracted from such curves is the substrate concentration giving the half-maximal turnover rate (called the Michaelis–Menten constant, or K_M), which is considered to be the substrate-binding affinity of the enzyme and identical for all of the enzyme molecules. This may be true, but alternatively it may only be accurate for the median enzyme in the population, as the K_M is based on the 50% level of activity. What if individual enzyme molecules vary in activity in a normal distribution across substrate concentrations (as they do; see Figs. 3.5 and 3.6)? This could also occur for sensor sensitivities, where subfractions of the sensor population would be triggered at low signal levels, and their own signals would be saturated, but if the signal goes higher, more sensors would be recruited and magnify the response (see Chapter 6).

These examples indicate that the possibility of variation among individual entities at the molecular level has been largely ignored during the development of biochemistry and molecular biology. This is due in part to the absence of methods for measuring individuals at that scale, so all such measurements effectively are means pooled across whatever populations may be present in a cell or tissue extract. This situation has now changed, with single-cell and single-molecule techniques available and being rapidly exploited (Liebherr et al., 2014; Emani et al., 2024). It is beyond the scope of this book to attempt a revision of all of molecular biology to PBT principles, but I hope that it might stimulate others in molecular biology and biochemistry to look again at their basic data, and ask whether they see an underlying population basis to the observed patterns (cf., Beal, 2017). If so, understanding how such variation contributes to actual biological functioning at its own spatial scale, rather than dealing only with means derived from pooled samples, would seem to be a key target for future research.

Response kinetics

The second key feature of the PBT model is related to the kinetics of the response triggered by the factor level. It turns out that the time required for the response to occur once the threshold is exceeded is inversely proportional to the amount by which the factor level exceeds the threshold, or the value of $X–X_b(i)$. This can be shown as:

$$\theta_X = [X - X_b(i)]\, t_i \tag{2.1}$$

where θ_X is a time constant for response to factor X and t_i is the time to respond for sensitivity fraction (or individual) i in the population threshold distribution ($X_b(i)$). θ_X is a constant, so as $X–X_b(i)$ increases, t_i (time to response) decreases proportionately. Thus, as the factor level increases above the response threshold, the response time is decreased in inverse proportion. This equation can be rearranged in terms of the rate of response ($1/t_i$) as:

$$Response\ rate = 1/t_i = [X - X_b(i)]\,/\theta_X \tag{2.2}$$

In essence, this is the population-based threshold (PBT) model. The population basis is embodied in the normal (or other) distribution of thresholds among the individuals [$X_b(i)$], the factor level [X] determines what fraction of the population is triggered to respond, and among those individuals that are triggered, how fast they will respond is based on the extent by which the factor level exceeds their threshold and the value of the time constant [θ_X]. How this works is illustrated in Figure 2.4. Specific examples will be given in subsequent chapters to illustrate how this is consistent with data from diverse biological systems.

Note that the strongly right-skewed response time courses (Fig. 2.4B) become more symmetrical sigmoid cumulative curves with decreasing maxima (asymptotes) when plotted on a log timescale (Fig. 2.4C; Dahal et al., 1990; Dahal and Bradford, 1994). This is consistent with the log-normal patterns proposed to be the natural outcome of the biological complexity of the processes associated with gene expression (Beal, 2017, pp. 57, 60):

> Consider that the operation of nearly every biochemical reaction involved in determining genetic expression should be expected to vary at least slightly from one cell to another due to differences in the state of individual cells, such as their size, health, available pools of various resources, and so on. This might seem to indicate that it is hopeless to attempt to model cell-to-cell variation without a full accounting of the expected distributions of each of these many models. That very complexity, however, leads to an emergent property [convergence to log-normal distributions] that provides a different route to abstraction with a strong mechanistic and mathematical grounding. . . . Log-normal convergence may also be applicable to a wide variety of other complex biochemical processes structured as catalytic cascades. Finally, it may be worth noting that in this case the complexity of biological systems results in simplicity rather than intractability, and that taking such aggregate-based perspectives may turn out to be valuable in dealing with other instances of biological complexity as well.

The PBT equation also indicates that the same response will be generated if the threshold distribution shifts higher or lower relative to a constant value of X just as if the value of X were shifted by the same amount in the opposite direction. That is, only the difference between X and $X_b(i)$ is relevant to the response, not their actual values. This is also important in physiology, where response thresholds can change because of other interacting signals or transitions to different developmental stages. In a physiological context, the simplest way to shift threshold distributions might be to increase or decrease the expression (or activation state) of a given receptor, which would then

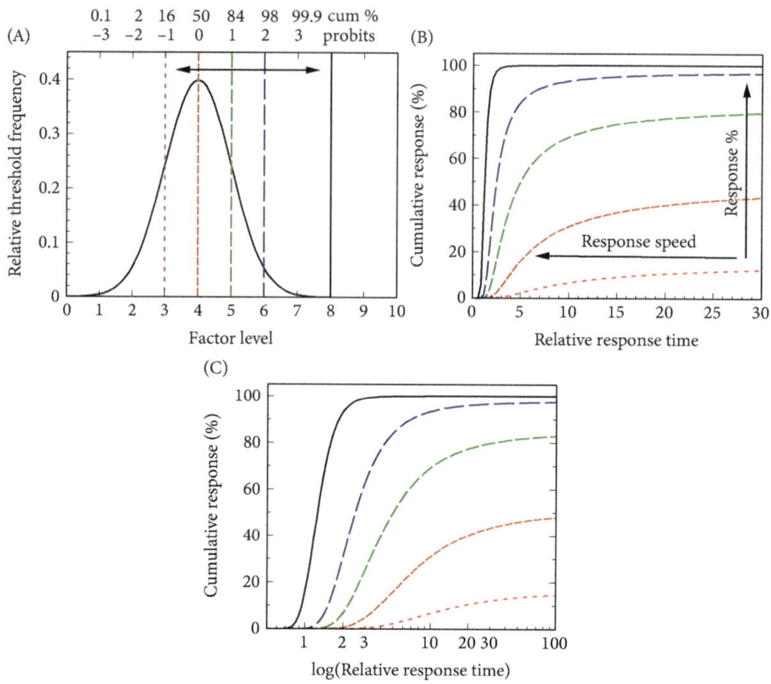

Figure 2.4 Illustration of the effects of different promotive factor levels on time courses of responses across a threshold sensitivity distribution. **A.** As the factor level changes (colored vertical lines), the response thresholds of different fractions of the population (black distribution) are exceeded, triggering them to respond. **B.** For each individual in the population, the speed of response is proportional to the amount by which the factor exceeds their threshold sensitivity. Thus, a high level of a promotive factor that exceeds all thresholds (e.g., level 8) results in rapid and complete responses of the entire population (black curve). However, at lower factor levels, both the fraction of respondents and the speed of their responses are reduced. Due to the variation in sensitivity thresholds within the population, there is a greater effect of factor level on the speed of slower fractions than on more rapid fractions, resulting in the skewness of the time courses. This is because the nearer the factor level is to the threshold of a given fraction, the time to response increases exponentially, reaching infinity (i.e., no response) when $X = X_b(i)$. That is, a change in factor level from 8 to 6 has a much greater effect on individuals with a sensitivity of 5 than on those with a sensitivity of 2. The time courses at lower factor levels will asymptotically approach the final percentages predicted on the scale above panel **A**; increasing factor levels will increase the response percentage (recruitment effect) and the speed of response for all responding individuals (kinetic effect). **C.** Plot of the response time courses in **B** on a log timescale. Due to the exponential delaying effect of the factor level approaching the threshold, plotting on a log timescale normalizes time courses to a more sigmoid shape and also illustrates the longer times required for all of the responses to occur at lower factor levels (i.e., to asymptote at the percentage corresponding to the fractions triggered in the normal sensitivity distribution).

adjust the sensitivity of downstream effects (e.g., transcriptional networks) to the factor level sensed by that receptor. Recent structural protein studies of ligand-receptor binding indicate that such properties can be affected by changes in amino acids throughout the receptor protein (not only in the binding region), suggesting a wide range of structural and physiological plasticity in these interactions (Heydenreich et al., 2023). The relative sensitivities of seeds to two hormones that either promote or inhibit their germination capacity change as the seeds enter or leave dormancy in response to temperature signals from their environment (Topham et al., 2017). In animals, individual neuronal pathways adjust their sensitivities to firing based upon prior experience and the signaling networks in which they operate: "This connectivity variability, or 'neurovariability,' is critical to the individuals we are" (Thiebaut de Schotten and Forkel, 2022, p. 507). Thus, the threshold distribution of a given signal-response is not a fixed property but is subject to adjustment in response to physiological, developmental, or environmental inputs.

The time constant also is not necessarily constant under all conditions. For microorganisms, plants and cold-blooded (poikilothermic) animals, the time constants must be adjusted for temperature effects on response rates using thermal time (degree-days), as is widely appreciated (Trudgill et al., 2005). The PBT model also enables normalization of the timescales of responses on the basis of other factors (e.g., hydrotime, hormone time) simply by multiplying the response times of each individual $t_i(X)$ by the value of $[1-(X/X_b(i))]$ (Bradford, 1990; Ni and Bradford, 1993; Dahal and Bradford, 1994; Alvarado and Bradford, 2005). Developmental events can also change the value of the time constant, which has proportional effects on the timing of responses across the sensitivity range. This normalized biological timescale has been termed "biotime" and is analogous to the use of "thermal time" (i.e., degree-days) to normalize the effects of different temperatures on rates of growth or development of cold-blooded (poikilothermic) organisms, but it also can be applied to the effects of hormones and other biological factors (Bradford and Trewavas, 1994). In addition, multiple factors can be combined in a PBT model to describe how the population responds collectively and simultaneously to them. For example, temperature and water potential (or water activity) are two major factors influencing seed germination, and PBT models can illustrate and predict their combined effects (i.e., "hydrothermal time") (Gummerson, 1986; Alvarado and Bradford, 2002). Similarly, promotive and inhibitory hormones influence seed germination timing; PBT models can be combined to predict the germination responses to different concentrations and combinations of these factors (Ni and Bradford, 1992; Ni and Bradford,

1993). Furthermore, as will be discussed frequently, what appears to be a single population distribution may in fact be composed of multiple subpopulations that can have distinct values for the parameters of the PBT model (Bello and Bradford, 2016; Bello, 2020; Bradford and Bello, 2022). In many cases, the existence of such subpopulations will be obscured by using pooled samples of the total mixed population (e.g., pooled cells in tissue or organ samples), highlighting the importance of the single-cell and single-molecule approaches currently being applied to appropriately understand underlying biological mechanisms (Leveau et al., 2018; Wagner et al., 2018; Mund et al., 2022).

I should note here that there are other modeling approaches, termed individual-based models (IBM) or agent-based models (ABM), that recognize the importance of individuality and agency in biology. These approaches have been applied particularly at higher levels of organization such as animal behavior and plant ecology (DeAngelis and Gross, 1992; Railsback and Grimm, 2019; Zhang and DeAngelis, 2020). These models share some key principles with the PBT model, such as the problems associated with the assumption "that all individuals in a population are identical and can be lumped together" (DeAngelis and Gross, 1992, p. xv):

> This assumption that many individuals may be aggregated into a single state variable transgresses two basic tenets of biology. The first tenet is that biological individuals are unique, differing from each other physiologically and behaviorally, depending on their specific genetic and environmental influences. The state variable approach may be viewed from either of two perspectives. In one view it assumes that all individuals represented by a population number are, for all practical purposes, identical. A second view is that there is no loss of information by considering just population averages. This view acknowledges that individual differences among organisms exist, but that they are not of concern to the issues under consideration, so that models which average over the full distribution of types in the population are sufficient.
>
> The second tenet transgressed is that of "locality," which means that a given organism is primarily affected only by other organisms in its spatial-temporal neighborhood. State variable models without spatial dependence implicitly assume that every member of the population has an equal influence on every other member of the population.

It is clear that the IBM/ABM are conceptually aligned with the PBT model, emphasizing the importance of individuality and agency in biology (Trewavas, 1999; Gilroy and Trewavas, 2023). IBM/ABM methods enable individuals to have distinct characteristics that can alter parameters contributing to the overall population level response. However, while such variation may have a stochastic basis, stochasticity is utilized primarily to add randomness to the parameters used in the modeling process, rather than as a basic structural feature as in the PBT model.

Conceptual metaphors for PBT models

While there is mathematical rigor to the PBT models, the goal here is to introduce its conceptual application across all levels of biology without relying too heavily on equations. It is not possible for me to do this in equal depth and formality across this scope of topics. Instead, my hope is to provide a population-based lens through which others can view their biological systems of interest. They will be better able to adapt PBT concepts to their own specialties and adopt appropriate analysis or modeling approaches. Thus, I will close this introduction to PBT models by providing a couple of conceptual metaphors that can give a feeling for how the model works without requiring deep mathematical expertise.

Hydraulic metaphor

The first of these metaphors is the hydraulic metaphor, in which the factor level [X] is represented as the height of water in a tank (Fig. 2.5). The distribution of thresholds [$X_b(i)$] is represented by a series of ports at different heights on the side of the tank that will allow the water to escape. As the water generates pressure in proportion to its height above each port, there will be a range of pressures related to the positions of the ports, which represents the values of X–$X_b(i)$. As the time parameter [θ_X] has a constant value, it is represented by a series of cups of the same size that will be filled individually by one of the ports. The time when a cup is full represents the time when the response will occur for that fraction of the port (sensitivity) distribution. Various scenarios can be envisioned using this scheme, such as when the factor level is lower (some ports stop flowing and all slow down), or if the position or distribution of the ports is shifted up or down. Changing the time constant would mean larger or smaller cups to be filled, thus adjusting the timing of all response times proportionately. These variations have been presented in an ecological context in Donohue et al. (2015), where further examples utilizing this metaphorical model can be found.

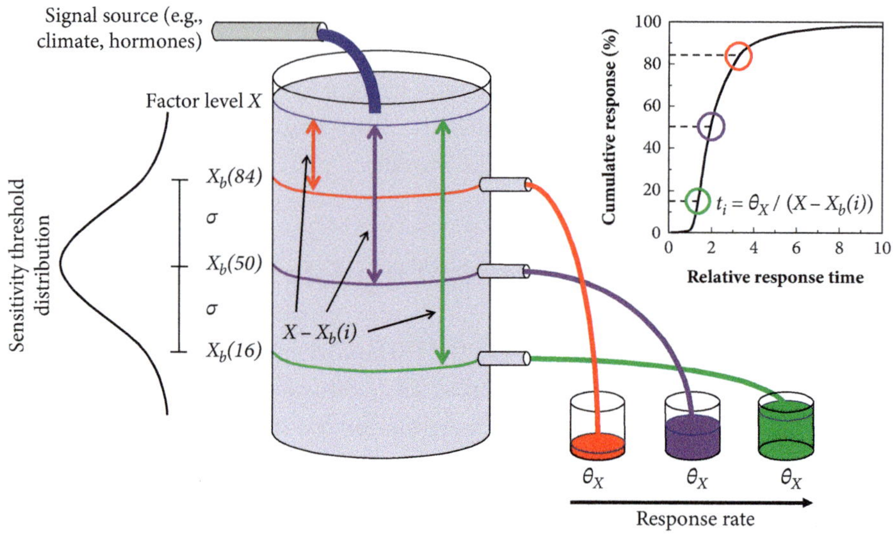

Figure 2.5 Hydraulic conceptual metaphor of the population-based threshold model. The factor level [*X*] is represented by the height of water in the tank, the threshold distribution [$X_b(i)$] by the ports in the side of the tank, and the time constant [θ_X] by the size of the cups being filled by the ports. The height of water above an individual port represents the factor level in excess of the threshold, or $X—X_b(i)$ (colored lines with arrows). A response occurs for a particular fraction when the respective cup is filled, as indicated on the response time course graph.

Adapted with permission from Donohue, K., et al. (2015). Applying developmental threshold models to evolutionary ecology. *Trends in Ecology & Evolution* 30: 66–77. http://doi.org/10.1016/j.tree.2014.11.008.

Rolling ball metaphor

We also have developed a second metaphorical model, termed the rolling ball model (Bradford and Bello, 2022). This model also illustrates the basic features of the PBT model using an alternative approach that emphasizes the response rates (Fig. 2.6). The model allows balls to roll down an incline and over subsequent inclines of different heights representing the response thresholds, which vary in a normal distribution. The distances that the balls roll after topping the incline are related to the height of the threshold "hill"; that is, balls topping higher hills (using more of their "momentum") roll shorter distances afterward, indicating their slower response rates. The slopes of the initial inclines relate to the values of the time constant (θ_X), with steeper slopes indicating smaller values and more rapid response rates. The heights of the hills represent the thresholds; if the starting point on the incline (factor level) is lower than the height of a given hill (response threshold), the ball will be unable to top the hill, indicating no response.

These metaphorical PBT models have been presented in these examples as a single normally distributed population (e.g., Fig. 2.3). However, while that is the case for an individual population (which also can change its sensitivity developmentally or in response to conditions), responding populations (e.g., enzymes, cells, etc.) also can be subdivided into multiple underlying subpopulations, each of which may vary in their signature parameters (mean, standard deviation, time constant). Just as what are considered to be analog responses of uniform populations are often quantal or digital (on/off) responses of individuals that appear to be analog because of recruitment of additional individuals as factor levels increase (see Fig. 4.4), changes in responses with effectors or over developmental time can also be due to shifts in the threshold levels or recruitment (or exclusion) of subpopulations with differing response thresholds. Multiple cases throughout this book will illustrate how apparently uniform biological responses could actually be the combined responses of multiple subpopulations that can be distinguished as methods increasingly focus on individuals rather than on pooled samples (see, e.g., Figs. 4.4, 5.6, 5.9, 5.12, 6.1, 6.8). This can be accommodated into the hydraulic model above by, for example, attaching additional groups of ports into the tank of the hydraulic model at higher or lower levels that will therefore be active (or not) as determined by the water level in the tank (i.e., the inducing factor level) (Fig. 2.5). In the rolling ball model (Fig. 2.6), additional subpopulations would be accommodated by additional routes for the balls to follow to reach the bottom of their incline and/or directing them onto inclines of different slopes and heights. That is, the same initial factor level can result in diverse responses depending upon the distribution of "routes" that the individual balls can encounter, corresponding to different subpopulations that might respond distinctly to that factor level.

Scope of application of PBT approaches

An additional advantage of the PBT concept is that it is applicable across all scales of biology. As will be documented in the following chapters, there are examples from the initial formative stages of biology in the (pre-) RNA world all the way to ecosystems where the PBT approach can provide unique insight into biological mechanisms. Each level of biological complexity can be viewed in a population-based context, with individuals at a lower level expressing

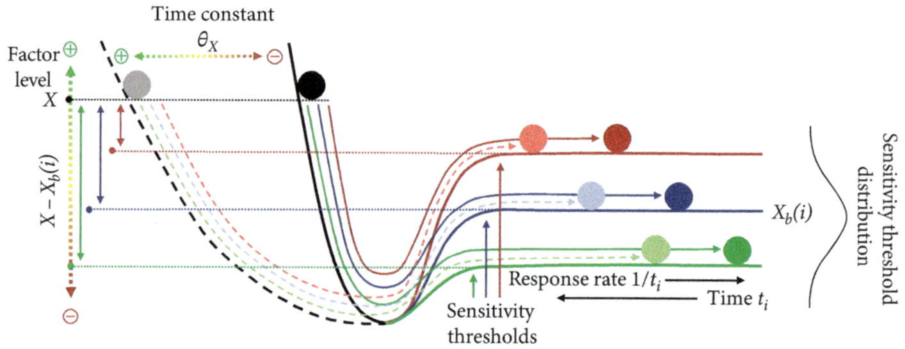

Figure 2.6 Rolling ball metaphorical model of the population-based threshold model. Balls can be released on an incline at different heights, corresponding to the levels of a promotive factor *X*. (For an inhibitory factor, the direction of the effect on the far-left scale is reversed; i.e., increasing inhibitory factor levels are associated with lower starting points for the balls.) The slope of the incline corresponds to the time constant [θ_X], with a steeper slope indicating a smaller value (dark balls, solid lines) and a shallower slope indicating larger values (shaded balls, dashed lines). After descending, the balls must roll up and over a hill, corresponding to the sensitivity thresholds, which vary in a normal distribution. The distances that the balls roll after the hill indicate their response rates [$1/t_i$].

Adapted with permission from Bradford, K.J. and Bello, P. (2023). Understanding seed behavior: populations of individuals. *Acta Horticulturae* 1365: 1–16. https://doi.org/10.17660/ActaHortic.2023.1365.1 and further detailed in Bradford and Bello (2022) (Open Access). An animated version can be viewed at https://pbt-models.github.io/pbtm-site/.

outcomes that provide the basis for (or input into) the next higher level of organization (Fig. 2.7). The nucleic acids and proteins associated with chromatin and gene expression provide the repositories of biological information and ways to access them, which are then used by cells to find resources, grow and multiply, eventually forming multicellular tissues, organs and organisms. Each of these levels is itself composed of populations of individuals, whose functions and sensitivities to internal and external inputs enable the whole system to function efficiently. The entire project of biology is to reverse the inexorable trend toward increasing entropy, and the biosphere represents an enormous case of negative entropy, or increase in complexity, powered by the ability to generate energy and maintain organizational continuity and metabolic homeostasis despite the ubiquity of random or stochastic variation (Kauffman, 2019; Kauffman and Roli, 2023). How biology managed to live with and coopt such random "noise" or "instability" will be a subtheme throughout this book.

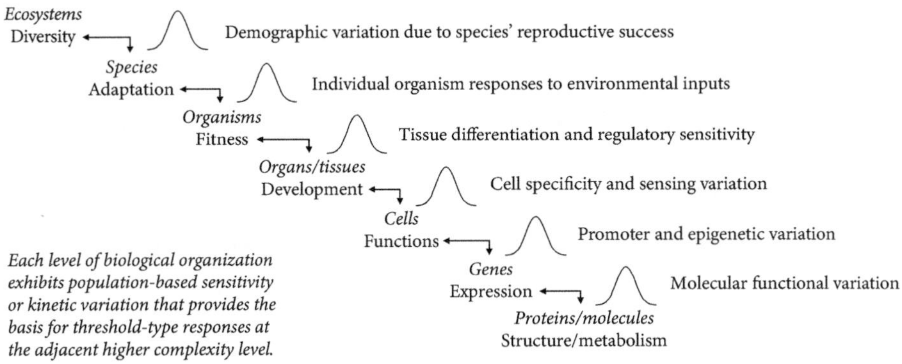

Figure 2.7 Organizational hierarchies of biological systems that function as populations with sensitivity threshold variations that underlie the increasing complexity at adjacent organizational levels.

Reproduced with permission from Bradford, K.J. (2018) Interpreting biological variation: seeds, populations and sensitivity thresholds. *Seed Science Research* 28(3): 158–167. https://doi.org/10.1017/S0960258518000156.

Conclusions

With this introduction to the basic PBT model, we are now ready to look at whether it seems to fit with what is observed in the biological world. Although the PBT model was first developed and applied to describe seed germination behavior (Gummerson, 1986; Bradford, 1990), and the most extensive examples have been demonstrated there (Bradford and Bello, 2022), it makes more sense here to follow an evolutionary pathway. Thus, the next chapter will start with molecules, nucleic acids and proteins before progressing to microorganisms, organelles, cell biology and organisms in subsequent chapters. At the tissue and organism levels, we can bring in seeds and other plant systems and then similarly examine animal biology before extending to development, evolution and ecology. So put on your population-colored glasses, and let's discover what we can see.

3

Origin and Integration of Biological Processes

Variation and the origin of life

To begin at the beginning, what was the situation in which biology first arose about 3.7 billion years ago? A full discussion of the origin of life ("abiogenesis") is beyond the scope here, and there are still diverse viewpoints on the topic. However, in thinking about how reproducing entities could arise in a purely chemical environment, Kauffman (1993) argued that self-organization and selection would be essential components. It is known that critical building blocks of life were available through abiotic chemical processes and that at least some of these, including amino acids, nucleic acids and membrane components, can arise through spontaneous chemical processes (Orgel, 2004; Szostak, 2017) and also arrived on Earth via meteorites (Kvenvolden et al., 1970). An essential key to life was the ability of some of these molecules to develop the capacity to reproduce themselves, giving them a selective advantage to increase their numbers compared to random chemical processes. As Nikolaos Papastavrou (first author of Papastavrou et al., 2024) stated, "We've long wondered how simple life was at its beginning and when it gained the ability to start improving itself. This study suggests the dawn of evolution could have been very early and very simple. Something at the level of individual molecules could sustain Darwinian evolution, and that might have been the spark that allowed life to become more complex, going from molecules to cells to multicellular organisms" (quoted in TechnologyNetworks, 2024). Or perhaps groups of molecules that could reproduce the members of the collective ("small-molecule collectively autocatalytic sets") could form once a threshold of overall efficiency was passed (Kauffman, 2019; Kauffman and Roli, 2023), likely aided by enclosure in lipid droplets and cyclical wetting and drying (Damer and Deamer, 2015).

For our purposes here, we can imagine that initially these processes were barely above the random threshold or were poorly reproducible. The processes of reproduction in biology are not exact, with variants occurring

Populations of Individuals. Kent J. Bradford, Oxford University Press. © Kent J. Bradford (2026).
DOI: 10.1093/9780197846032.003.0003

continuously due to spontaneous mutations or errors in the reproductive processes associated with the synthesis or duplication of nucleic acids (Aebersold et al., 2018), even with current error-checking and repair processes that reduce the error rate by a million-fold (Noble and Noble, 2018). We can imagine, therefore, that the original chemical "soups" would have been occupied not by specific molecules precisely reproducing themselves, but rather by networks of diverse related structures, created by chemical processes (including primitive catalysis), with the collective ability to enable their own self-reproduction (Kauffman and Roli, 2023). Those collectives that did attain the ability to reproduce themselves immediately gained dramatic advantages in increasing their numbers relative to others relying only on random chemistry (Szostak et al., 2001). While nucleic acids do not seem to be essential for such initial self-reproducing sets (Xavier et al., 2020), RNA molecules would have become involved at some point early in this chemical evolution into self-reproducing entities, as they are essentially ubiquitous in biology now and still have the ability to reproduce copies of themselves and to carry information (Kim et al., 2021). The remarkable variety of RNA forms and functions that we are currently still discovering, and their occurrence and preservation across all levels of biological organization, suggest that the "RNA world" played at least some part in the early expansion of biology (Assmann et al., 2023).

As hard as it may be to comprehend, biology originated from these widely diverse and poorly reproducible processes. As membranes and compartmentation developed, creating the possibility for somewhat more contained and advantageous environments in which these processes could occur, that is, allowing individuation of these autocatalytic collectives (Kauffman, 2019), we have to imagine that the underlying processes would have gradually become more efficient and directed. In fact, an increase in the fidelity of replication of an RNA polymerase ribozyme due to random mutations is associated with a major increase in reproductive fitness (Papastavrou et al., 2024, pp. 1, 2, 8, 9):

> As the efficiency and accuracy of RNA replication improved, larger and more complex RNAs could be replicated, encompassing more sophisticated catalytic motifs and expanding the functional repertoire of the RNA world. Throughout the generations of RNA-based evolution, copying accuracy must have exceeded a critical threshold to maintain heritable information, and this threshold would have risen as the evolving RNAs increased in size and complexity. . . . These results are relevant to the RNA world hypothesis, whereby the higher the fidelity of the polymerase, the larger the functional RNA it is able to synthesize; and the larger the functional RNA it is able to synthesize, the more opportunity there is to evolve complex motifs with even greater fidelity.

> Given the high mutation rates . . ., the object of selection is not a single peak sequence, but rather a constellation of variants that include the peak sequence and closely related sequences that evolve together as a quasispecies. . . . The structure of a quasispecies is determined both by the available pathways for mutational interconversion among its members and by the underlying fitness distribution of those members. . . . [T]he higher-fidelity polymerase enables an evolutionary progression of successive peak sequences and corresponding quasispecies that move progressively further from the starting sequence, while remaining within the central portion of the contour map for . . . functionality.

Note that the authors have included the concept of thresholds in their description, in this case thresholds of polymerase fidelity that are sufficient to enable the maintenance of specific sequences over generations and to prevent dissipation of functionality due to random incorporation of errors over time. At the same time, while the range of variation is decreased, there is still a distribution of functionality among the remaining variants. This selective pressure no doubt continued, resulting in the sophisticated level of error detection, proofreading, correction and degradation currently embodied in cells to maintain their DNA, RNA and protein integrity and function (Komander and Rape, 2012; Waterworth et al., 2016; Bębenek and Ziuzia-Graczyk, 2018; Nickoloff et al., 2021; Eckardt et al., 2024). At the same time, the presence of these error-minimizing and self-correcting mechanisms in living organisms today indicates that the underlying potential for variation due to stochastic processes is still present. In fact, it has been argued that stochasticity is essential to enable choice or agency, which is evident to varying extents across all of biology (Noble and Noble, 2018; Kull, 2022). In addition, as increasing numbers of interconnected subprocesses became involved in early biology, each of which would bring their own fidelity threshold distributions, the combined likelihoods would result in something resembling a normal distribution across the entire population, reflecting their still fundamentally stochastic basis. This leads to the conclusion that since variation was inescapable at the origin of life, early protocells and the initial single-celled organisms must have arisen within that context and developed the capacity to cope with it.

As there could not have been sophisticated error-checking and correction processes at a time when error itself meant only a slightly lower probability of replication among barely more than random entities, those entities that did propagate themselves would have had to adapt their systems to deal with this inherent variability. Individual entities that could maintain and reproduce themselves did develop under these conditions, no doubt incorporating much random variation along with their elementary levels of control and reproducibility. In fact, emerging organisms actually "harnessed" stochasticity as part of their fundamental functional mechanisms (Noble and Noble, 2018). As we will see shortly, current single-celled microorganisms contain remarkably sophisticated systems for basic functions that are largely retained across all biology. After four billion years of chemical and biological evolution, inherent variation along the lines described by the PBT model is still present and utilized by all organisms. It is also a truism in evolution that once something fundamental works, it is seldom completely abandoned, as evidenced by the legacy in eukaryotes of their archaeal precursors and bacterial endosymbionts (mitochondria and chloroplasts) (Ku et al., 2015; Zaremba-Niedzwiedzka et al., 2017). Thus, I take it as a highly likely premise here that the original organisms to evolve from the purely chemical world had to cope with inherent variation in their component parts. The fact that they did so includes the possibility that such variation is not only compatible with biological systems but may in fact be a fundamental feature of them.

Promoters, transcription factors, and gene regulation

Jumping ahead to what we now know about molecular biology in all currently existing organisms, it is fundamentally based on the abilities of nucleic acids to reproduce themselves and to coordinate the synthesis of RNAs and proteins that carry out the essential functions of life. The systems of genome management, that is, the care and feeding of DNA molecules, are now highly complex, even in single-celled organisms. Breaking this system down to its essentials, we find that the DNA contains regions (genes) with specific nucleotide sequences that code for the amino acid sequences in the proteins required for the organism to live. The process of converting those nucleotide sequences into proteins has two parts: the transfer of the information embedded in the DNA nucleotide sequence into a corresponding sequence in a messenger RNA (mRNA) molecule in the nucleus, and the use of that mRNA molecule by ribosomes in the cytoplasm to direct the synthesis of the specific protein it encodes. The first process is coordinated by a system of regulatory sequences (promoters) adjacent to the coding sequences in each gene. Specific proteins (transcription factors) can recognize unique signal sequences in the promoters, bind to them, and attract other proteins, particularly RNA polymerases, that transcribe the DNA sequence into a newly synthesized complementary mRNA molecule. A given transcription factor may be able to recognize multiple promoters (or other locations within the coding sequence) and to control the production of other transcription factors, in a cascading feedback regulatory network. This is particularly evident and critical during development in multicellular organisms, as the transcription factors are key players in the differentiation of diverse cell types and the production of specific proteins and functions associated with distinct cell types. However, they are also critical for maintaining the essential functions required for life itself, including acquiring food, generating energy, maintaining cellular integrity and performing specific tasks such as reproduction or resisting disease organisms.

This brief overview of basic molecular biology is well known to all biologists. What I want to do now is to cast this system within the PBT structure described previously. It is obviously a very complex system, with each cell in a human containing the genes coding for about 30,000 proteins. This number increases to nearly 100,000 in wheat, each cell of which contains three complete (but significantly different) genomes (Shi and Ling, 2018). The primary interaction enabling transcription is the binding of specific proteins to correspondingly specific short nucleotide sequences in the promoter regions of the genes (Fig. 3.1). This is the critical interaction that allows specific transcription factors (TFs) to control only a unique fraction of the total genes present. Thus, we can think of a situation within the nucleus in which there are populations of regulatory proteins (e.g., repressors [Rs] or enhancers [Es]) and TFs, each potentially at a uniquely determined concentration, along with a diverse population of promoters encoding sites to which these proteins can specifically bind.

As shown in scenario 1 in Figure 3.1, repressors can bind to the promoter and prevent the binding of TFs, which also prevents transcription of the adjacent gene. Repression can also be achieved by "epigenetic" processes such as DNA methylation or histone protein modification, which have their own binding sequences and ancillary proteins that guide them to the correct sites. Interestingly for our discussion here, "silencing and reactivation occurred in all-or-none events, enabling the regulators to modulate the fraction of cells silenced rather than the amount of gene expression" (Bintu et al., 2016, p. 720). At the relevant molecular scale, the interactions among nucleic acids and proteins will be governed primarily by ionic and physical interactions dependent upon their molecular (and three-dimensional) structures, diffusion rates, concentrations of the proteins, and related biophysical principles (Parthasarathy, 2022). From a PBT viewpoint, gene repression removes a gene from the potentially expressible subpopulation of genes in the genome. That is, the sensitivity of those genes to TF binding is very low or nonexistent.

In scenario 2 of Figure 3.1, the repression has been removed by lowering the concentration of the repressors, demethylation, histone modification and so on. It is now possible for TFs to potentially bind to the subpopulation of derepressed genes that have the appropriate recognition sequences in their promoters. Even then, these binding sequences can often tolerate some variations in their nucleotide sequences. Thus, the population of promoters in the genome varies in the degrees to which their binding sites match the optimal expectations of individual proteins in a population of TFs. In scenario 2, the match between TF1 and its promoter binding site is sufficient to allow binding, which is then followed by binding of additional proteins that enable

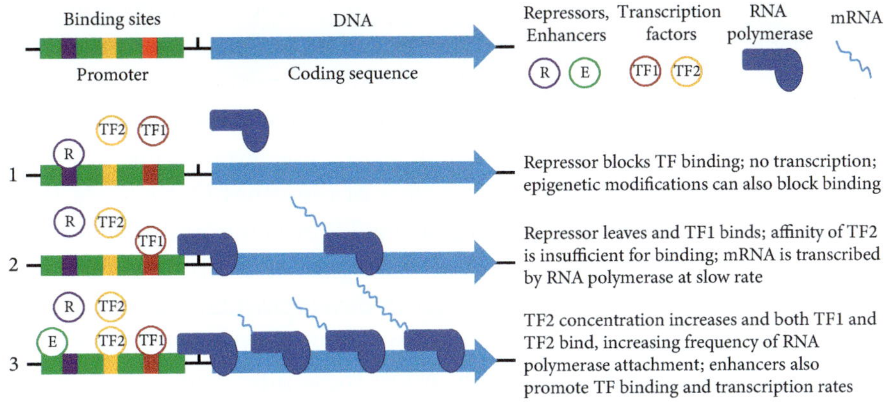

Figure 3.1 Schematic diagram of key components of gene transcriptional regulation. A gene is shown with a promoter and coding region. Binding sites for repressors (R) or transcription factors (TF1, TF2) are indicated. Symbols also indicate RNA polymerase and an mRNA molecule being synthesized. (1) The first scenario indicates that the gene is repressed by a specific repressor protein (R), or by other methods such as DNA methylation or histone modification. RNA polymerase is neither attracted nor bound, and transcription does not occur. (2) The second scenario indicates that when repression has been released, a TF can bind to its specific site and to an RNA polymerase (and multiple associated proteins) and initiate mRNA synthesis. The RNA polymerase complex then progresses down the length of the coding region, constructing a mRNA molecule encoding the (complementary) DNA nucleotide sequence. (3) The third scenario indicates the possibility of multiple TFs binding to a given promoter, potentially enabling more rapid attachment of RNA polymerases, thereby increasing the rate of mRNA transcription. Enhancers (E) encoded in DNA and brought near the promoter via DNA structural configurations can also promote TF binding. The possibility of regulation by multiple TFs, Rs, and Es permits an individual gene to be involved in multiple regulatory networks. See text for description of how these scenarios can be viewed within the PBT concept.

the attachment of RNA polymerase to the transcription start site and the initiation of transcription. However, in this scenario, TF2 is present at a concentration below its binding threshold to the promoter site, so only TF1 can enable transcription.

In scenario 3, the concentration of TF2 is increased, allowing it to bind to its lower sensitivity receptor site on the promoter. Alternatively, enhancers encoded in nearby DNA regions can bind to the promoter region and maintain the DNA in its "open" state, allowing additional transcription factors to bind (Dey et al., 2015; Bohrer and Larson, 2021). This enhances RNA polymerase binding to this promoter and thereby increases the rate of transcription of the gene. Thus, it is possible that a promoter might interact with multiple subpopulations of TFs, Rs and Es, and that their combined action regulates the transcription frequency of its associated gene sequence. It is not uncommon for a gene to be part of a regulatory network that is responsive to multiple inputs, and for transcription to exhibit "bursts" rather than a consistent rate (Finn and Misteli, 2019). The levels of the TFs themselves are also regulated by the promoters of the genes that encode them, and by sequestration or degradation of the TF proteins, and therefore can be sensitive to regulation by diverse physiological or environmental inputs. In many cases, particularly with respect to developmental changes, a given TF can regulate the transcription of multiple downstream TFs, greatly multiplying its effects on gene expression. All of these interactions make it easy to understand how the final transcriptional output of a specific cell at a specific time might be considered to be "noisy" (Raser and O'Shea, 2005; Tsimring, 2014; Harton and Batchelor, 2017; Ham et al., 2021), particularly when thousands of genes may be in the process of being transcribed at the same time.

However, when looked at from the PBT viewpoint of an individual promoter, its sensitivity to binding a given TF is encoded in its sequence. As the levels of different TFs, Rs and Es change in response to development, energy levels, stress or environmental signals, the individual genes will simply respond in a quantal (on/off) manner to the TFs that successfully bind to them. TFs themselves are often developmentally regulated or respond to diseases or stresses. In this case, they are the factors whose levels vary in the PBT models, while the promoters and their associated coding regions are the responding population that is sensitive to the TFs (cf. Fig. 3.1). If the TF concentrations increase to high levels, they may successfully bind to promoters whose sites are not optimal (i.e., with a lower sensitivity), but may still enable binding at a higher TF level. Thus, the derepressed promoters represent a population with a distribution of sensitivity thresholds that can respond to corresponding concentrations of TF populations. In addition to this "molecular switch" type of TF binding to promoter sites, "allosteric" and "multivalent" activation is also possible (Lauzon and Vallée-Bélisle, 2023). An allosteric activator can increase the affinity of the promoter site for a TF, effectively increasing the sensitivity to it. Similarly, a multivalent activator can bind to a site and change its conformation to reveal an additional hybridization interface and enhance TF binding to the site, again increasing sensitivity to the TF. These mechanisms can be distinguished by the ways that the dose-response curves shift in the presence or absence of the activators. In addition, "Multivalent activators also enable programing of the dynamic range of a switch independently from its apparent affinity, leading to precise control over the activation threshold and the sensitivity of molecular switches" (Lauzon and Vallée-Bélisle, 2023, p. 18852). Thus, these mechanisms illustrate that the mean threshold sensitivity and its variation for a given gene with respect to transcription can be efficiently modified (Harton and Batchelor, 2017), in keeping with the PBT model. In fact, it has been demonstrated that population distributions of binding behavior of individual molecules affecting DNA accessibility are sensitive to single-nucleotide mismatches (Aguirre Rivera et al., 2024). While exceedingly complex at the scale of even a single gene, the PBT concept makes it look much simpler from the point of view of the participants. Each individual promoter is simply responding in an on/off manner when its activation threshold is exceeded by one or more TFs binding to it, which is controlled by the promoter sequence, its three-dimensional configuration, and the activators that can promote or inhibit access of TFs to the transcription start site.

In addition, as indicated in scenario 3 of Figure 3.1, the passage of an RNA polymerase molecule down the DNA coding region can be considered to be a quantal event, as the rate of progress of the enzyme complex down the DNA sequence seems to be relatively fixed (Ham et al., 2020). Increasing the transcription rate of an individual gene therefore must involve initiating additional polymerase complexes more frequently, rather than speeding up their progress once started. When TFs exceed the minimum sensitivity (binding) threshold, the PBT model will predict that the transcription rate will increase; that is, the number of RNA polymerases progressing through transcription per unit time will increase. In addition, genes coding for proteins that are needed in large quantities at a given time often have multiple copies (alleles) of those genes in the genome, such as for genes encoding storage proteins that accumulate rapidly during seed development or heat-shock proteins that respond rapidly to high-temperature stress. Increasing the size of the population of such genes provides an automatic way to increase the rate of production of their products using the same basic transcriptional and regulatory mechanism.

In many cases, biological responses follow logarithmic dose-response concentration curves (e.g., Lauzon and Vallée-Bélisle, 2023). A logarithmic range of responses implies that some individuals may be orders of magnitude more sensitive to a given signal than others. This logarithmic variation in response thresholds is so common that it is seldom remarked on, but it is significant when interpreting the effect of a given concentration (see Fig. 5.5 or 6.2, for example) or comparing physiological concentrations of effectors with dose-response curves. Such variation is also evident in the timing of transcriptional bursts of gene expression. Larsson et al. (2019) reported that the timing between transcriptional bursts for different genes exhibited a right-skewed frequency diagram (Fig. 3.2A). However, when this burst timing is plotted on a logarithmic scale, it becomes a normal distribution (Fig. 3.2B), and plotting this curve over the frequency diagram on a linear scale shows how well this matches the data (black lines in Fig. 3.2A). A potential source of such logarithmic timing is the exponential increase in the time to response predicted by the PBT model as the factor level nears the sensitivity threshold (Fig. 2.4).

The underlying structure of DNA in chromatin complexes also contributes to the heterogeneity of transcription (Finn and Misteli, 2019), and there is

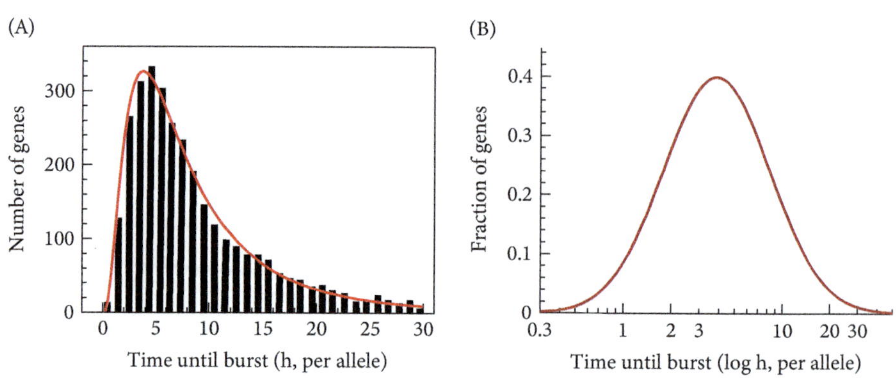

Figure 3.2 Timing of transcriptional bursts across many genes in mouse fibroblast cells. **A.** Frequency diagram (bars) of the number of genes exhibiting transcriptional bursts after specific times. The red line is the distribution in panel **B** plotted on the linear scale. **B.** A normal distribution of gene numbers versus a log timescale for transcriptional bursts. The normal distribution has a mean of log 3.6 h and a standard deviation of log 2.15 h. When this is plotted on a linear scale, it is the red line in panel **A.**

Data in panel **A** (without red curve) reproduced with permission from Larsson, A.J.M., et al. (2019). Genomic encoding of transcriptional burst kinetics. *Nature* 565(7738): 251–254. https://doi.org/10.1038/s41586-018-0836-1.

considerable cell-to-cell variation in chromatin structure even in cultured cells (Su et al., 2020). Su et al. (2020) were able to categorize genetic loci into two regions into which chromatin is partitioned in the nucleus with "A" compartments enriched for active chromatin and "B" regions containing relatively inactive chromatin (Fig. 3.3). They observed that transcriptional bursting frequency is much higher in the more active A region than in the less active B region (Fig. 3.3A), consistent with bursting frequency being an indication of transcriptional activity and with the logarithmic variation in bursting frequency shown in Figure 3.2. The burst sizes in the two regions, however, overlapped much more (Fig. 3.3B). This is consistent with the wide variation in bursting frequency shown in Figure 3.2 and suggests that bursting frequency is a major determinant of variation in transcription rates both within a given nucleus and across cell populations. Dey et al. (2015) had previously described transcriptional on/off regulation (bursting frequency) as contributing to individual variation and burst magnitude as correlated with population mean. This could be interpreted in PBT terms as there being a distribution of effectors that might turn on the gene in a bimodal manner, which vary and therefore constitute the bursting frequency component. Once turned on, the burst size might be determined by how much the signal (TF concentration) exceeds the threshold, or essentially the kinetic component of the PBT model. This would mean that the sensitivity threshold distributions determine the bursting frequencies, while the amount made per burst (burst size) would be the result of more frequent ON transitions due to higher concentrations of the effector(s) above the threshold. These hypotheses can now be tested using single-cell methods.

Chromatin loops and clusters form domains (referred to as topologically associating domains, or TADs) that influence the likelihood of transcription in those regions (e.g., the A and B compartments in Fig. 3.3). The interactions of these genomic regions form population distributions related to their minimum distance apart: "Biochemical methods such as high-throughput chromosome conformation capture (Hi-C) provide genome-wide maps of contact frequencies, a proxy for how often any given pair of loci are sufficiently close in space to be captured together" (Finn et al., 2019, p. 1502). The interaction–distance relationships for many pairs of loci are shown in Figure 3.4A. These distributions look remarkably similar to the time courses generated by the PBT model (see, e.g., Figs. 2.4B, 5.8E, F), suggesting that distance may be a proxy for time at the molecular level. In fact, the data of Finn et al. (2019) can be matched by the PBT model using a single Hi-C frequency distribution and varying the time constant at different Hi-C frequencies (Fig. 3.4B). This is consistent with the time factor being associated

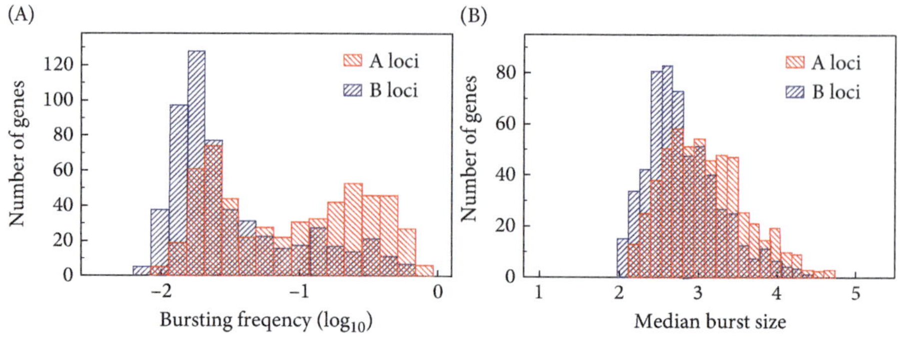

Figure 3.3 Distributions of transcription bursting frequencies (**A**) and burst sizes (**B**) for genes in compartments A (red: higher transcriptional activity) and B (blue: lower transcriptional activity) in the nucleus.

Reproduced with permission from Su, J.-H., et al. (2020). Genome-scale imaging of the 3D organization and transcriptional activity of chromatin. *Cell* 182(6): 1641–1659. https://doi.org/10.1016/j.cell.2020.07.032.

with the distance between interacting sites. These relationships also are heterogeneous across cells and can vary due to several mechanisms. Finn and Misteli (2019, pp. 4, 6) concluded the following:

> The stable and reproducible transcription profiles at the level of cell populations and tissues are thus consensus averages and represent the result of highly variable transcription of individual genes averaged over time and among all cells in the population. . . . The advent of single-cell sequencing and biochemistry methods and the use of quantitative imaging methods are enabling us to address this important aspect of genome function. Although variability in genome organization has in the past complicated attempts to understand the functional relevance of genome organization, the recent progress summarized here has made structural heterogeneity an experimentally tractable parameter and suggests that variability in genome organization is a driving force in genome function, rather than just a confounding factor in its analysis.

This conclusion is further supported by Jores et al. (2024), who showed that plant enhancers can also function both cooperatively and additively in conferring enhancer activity for light-responsive genes.

An additional mechanism for regulating transcription has been termed "epigenetic" regulation, which is based on modifications of the proteins that organize DNA structurally in chromosomes. These proteins, called "histones," can be modified to either prevent or promote uncoiling of the DNA and therefore control access to promoter sites. Various epigenetic "marks," including

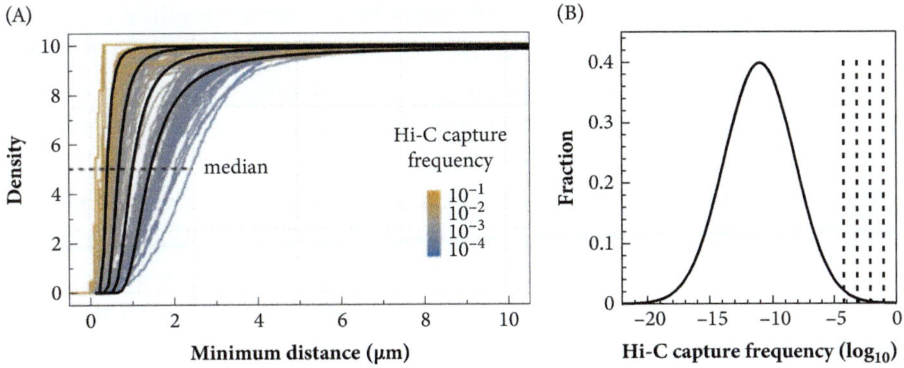

Figure 3.4 Genomic regions form population distributions related to their minimum distances apart. **A.** Cumulative distance distributions for pairs of interacting regions in the genomes of human fibroblast cells. "Hi-C capture frequency" refers to the propensity of two DNA regions to contact each other. The raw data in **A** are shaded from brown to blue from left to right. The black curves are the predicted distributions based on a common Hi-C frequency distribution (shown in panel **B**) but increasing the time constant as capture frequency decreases logarithmically. **B.** A population distribution for Hi-C capture frequencies. The values shown in panel **A** are indicated by the dotted lines. The best fits of the PBT model to the data were obtained when a common distribution but different time constant values were used (a mean of 10^{-11} and a standard deviation of 2.8 with time constants of 4, 6, 8 and 10 for frequencies of 10^{-1}, 10^{-2}, 10^{-3} and 10^{-4}).

Data in panel A are reproduced with permission from Finn, E.H., et al. (2019) Extensive heterogeneity and intrinsic variation in spatial genome organization. *Cell* 176(6): 1502–1515. https://doi.org/ 10. 1016/j.cell.2019.01.020.

DNA methylation, histone protein variants and histone post-translational modifications, influence the chromatin organization and availability of the DNA for transcription (Candela-Ferre et al., 2024). For example, a key gene (*FLC*) regulating the development of flowers in some plants in response to exposure to cold temperatures is regulated by epigenetic marks. As time in the cold increases (vernalization), individual cells switch to silencing *FLC* (which inhibits flowering), and this eventually reduces FLC protein to the point that the flowering developmental program can proceed. The authors' description of their model for *FLC* regulation illustrates its affinity to the PBT concept (Angel et al., 2011, p. 108): "Our combined experimental/modelling approach to vernalization has revealed that the quantitative level of epigenetic silencing in the warm is generated by an appropriate proportion of cells switching into a stable *FLC*-silenced state. Controlled switching of individual cells between bistable states, mediated by nucleation of an opposing histone mark may therefore be a common method of achieving a quantitative epigenetic response at a population level." (See Chapter 5 for further description of this system.) In addition, epigenetic marks can also promote initiation of transcription at multiple sites within a gene, including within the coding region, called alternative transcription start sites (ATSS) (Alfonso-Gonzalez and Hilgers, 2024). A recent study with soybean (*Glycine max*) found that approximately two-thirds of protein-encoding genes contained such sites, often having multiple sites per gene (Wang et al., 2024). The occurrence of such ATSS was associated more strongly with tissue- or development-specific genes, indicating the expanded potential for variation among the proteins expressed within a gene family or a transcriptional cluster.

Post-transcriptional control of mRNA abundances also involves "microRNAs." MicroRNAs (miRNAs) are short RNAs (~20 to 24 nucleotides long) that are transcribed as longer primary miRNAs encoded in DNA and then trimmed to their final size. They act in the cytoplasm to bind to mRNAs containing nucleotide sequences complementary to those in the miRNA, preventing translation of the mRNAs and promoting their degradation by RNases. They are also abundant (Ebert and Sharp, 2012, p. 515): "As a class, miRNAs constitute about 1%–2% of genes in worms, flies, and mammals (Bartel, 2009). Their regulatory potential is vast: more than 60% of protein-coding genes are computationally predicted as targets based on conserved base-pairing between the 3' UTR [untranslated region] and the 5' region of the miRNA (Friedman et al., 2009)." In addition, some miRNAs are conserved from sea urchins to humans (Mukherji et al., 2011), so it is also a very old mechanism in evolutionary history for regulating translational kinetics.

Of particular interest here is the potential stabilizing effect of miRNAs to offset some of the variation in transcription rates described above. The advent of single-cell analyses has further revealed the population basis of this function and its relationship to sensitivity thresholds (Mukherji et al., 2011, p. 854):

> Our single-cell analysis revealed two unexpected features. First, the experiments showed that regulation by miRNA imposes a previously unappreciated nonlinearity relating target transcript abundance to target protein abundance; namely, target protein production is highly repressed below a threshold level of target mRNA production and responds sensitively to transcription above this threshold. Second, there is considerable cell-to-cell variability in the strength of repression in a population of identically prepared cells. Most notably, the fold repression of miRNA targets below the threshold can be far greater than measured for the population average,

> up to 40-fold greater in our data. . . . [W]e employed and experimentally tested a mathematical model to formally describe the biochemical interactions comprising the miRNA regulatory system. The model suggested the importance of molecular titration in generating the sensitive response to transcription above the threshold. In this picture, as target abundance increases, target mRNAs titrate away the pool of miRNAs available for repression; the sharpness of the switch from full repression to escape from miRNA repression depends on the strength of the interaction between the miRNA and its target as well as the relative abundance of each.

This is essentially a description of how the PBT model works, with the sensitivity of binding between receptors (target mRNAs in this case) and their inhibitors (miRNAs in this case) determining a sensitivity threshold distribution along with exponentially decreasing responses as the inhibitory factor increases (see Fig. 5.8B, D, F). Also consistent with the PBT concept is the cellular variation in these interactions, illustrating the population basis of the regulatory system at both the molecular and cellular levels of organization.

In addition, miRNAs can serve to dampen the "noise" associated with transcription and increase the robustness of the system in the presence of changing sensory inputs (Wu et al., 2009; Posadas and Carthew, 2014; Schmiedel et al., 2015; see also Chapter 7). Part of this "noise" is associated with the switch-like behavior at the base of the system: individuals in the population function primarily in on/off mode, which is hidden when observations are only made on pooled samples or averaged over many cells. This has been explored in detail with respect to miRNA regulation by Siciliano et al. (2013, p. 5, 6):

> A PFL [positive feedback loop] enables a gene regulatory network to encode a "cellular memory" because of its switch-like behaviour created by two different stable states. Transitions across the states (that is, switching) can be triggered by a specific endogenous or exogenous event, thus enabling the control of key decision processes such as cell differentiation and cell fate in response to specific clues.
>
> The switch-like behaviour of the PFL motif, however, comes at a cost: first, the PFL may considerably slow down the transcriptional response of the circuit to the triggering event; this can be acceptable, or necessary, in some cases, but detrimental in others; the second and biggest issue is that the PFL motif increases biological noise because small fluctuations in the expression levels of the TF [transcription factor] are amplified by the self-regulatory loop. The increased cell-to-cell variability in the TF concentration may cause the TF level to be higher than the promoter activation threshold, inducing transition of the switch from the OFF state to the ON state, even in the absence of the triggering event.

> Here we experimentally demonstrated that the presence of a miRNA suffices to solve both issues at once, by conferring phenotypic robustness to the toggle-switch, thanks to the reduction of biological noise, and by enabling the circuit to respond faster. . . .
>
> Our results add further support to the role of miRNAs as noise suppressor. In addition, we demonstrated that reducing noise is essential to maintain a robust phenotype by preventing cells from unwanted stochastic switching.

This illustrates that feedback relationships can alter the transcription rates of the genes encoding TFs as their concentrations vary among cells during development. Interestingly, in this case both interacting factors (i.e., the target mRNA population and the miRNA population) can change in abundance based upon their own relative transcription rates in response to developmental or environmental inputs. That is, an increase in the miRNA population (i.e., higher concentration) relative to the mRNA population will repress translation, but a relative increase in the mRNA population can also saturate the existing miRNA pool and switch translation on (see also Fig. 5.3).

Variation can also occur among the translated TF proteins in relation to development. Papadopoulos et al. (2019) demonstrated variation in concentration and in physical characteristics (e.g., diffusion rates) among individual cells expressing various TFs in developing *Drosophila* imaginal discs. They showed that this cell-to-cell variation changed during development, with higher variation when fewer cells were expressing the TFs than later when more cells were expressing them. This was shown to be consistent with a population-based model in which a low number of cells expressing stochastic variation would exhibit greater apparent variation than when a large number of cells were expressing; that is, expression levels near the mean would dominate in the population when the overall protein concentration was high. This change in variation was also associated with a feedback mechanism for one of these TFs (Hox protein Antennapedia, or Antp) (Papadoupoulos et al., 2019, p. 6):

> [T]hese results revealed the existence of a switch in Antp auto-regulatory capacity on its own transcription during development. Starting from a preferentially auto-activating state early in development, Antp changes to an auto-inhibitory mode at later developmental stages. . . . [O]ur measurements show that Antp is expressed at relatively low and highly variable levels in early developing discs, when genetic evidence indicates auto-activation capacity on its own transcription. Later in development, when Antp has reached a state of higher average concentrations, auto-repression kicks in, resulting in considerably lower variability among neighboring cells.

Proteins

While some RNA molecules have regulatory effects on their own (e.g., miRNAs) (Assmann et al., 2023), in general the transcribed mRNAs are moved from the nucleus to the cytoplasm and attach to ribosomes that synthesize proteins from their amino acid building blocks based upon the nucleotide sequences coded in the mRNAs. This step is also based on extremely complex and interesting molecular biology at the single-cell level (Taniguchi et al., 2010), but we will focus here on only one critical aspect with major implications for the PBT concept. While some minor errors are inevitable, it is generally assumed that since the amino acid sequence is encoded in the mRNA, the proteins produced using a given mRNA template will have identical functional activities. However, a remarkable study by Raphaela Liebherr, Max Renner and Hans Gorris (Liebherr et al., 2014) found something different. They used bacteria to synthesize proteins of an enzyme (β-glucuronidase, or GUS) that converts a colorless substrate into a colored product that can be measured to indicate the enzyme's catalytic activity. However, rather than doing these assays in a test tube with millions of (theoretically identical) protein molecules, these researchers diluted the enzyme proteins to the extent that they could place only one individual molecule in a single microscopic "vial" holding only a femtoliter (10^{-15} liter) of assay solution. They then measured the activity of these individual enzyme molecules across a range of substrate concentrations, in accordance with the standard Michaelis–Menten kinetics approach (Nelson et al., 2008) in which activities (substrate conversion rates) at different substrate concentrations are measured and used to define the characteristics of the enzyme, such as reaction velocity and the substrate concentration at which this velocity is half-maximal (K_M). When measured individually and averaged, the researchers found that "their mean activity is consistent with classic Michaelis—Menten kinetics." However, when the results for each enzyme molecule were plotted individually (as subfractions of the total population), a very different picture appeared: the individual molecules did not have identical activities at a given substrate concentration. Rather, at each concentration, the mean activity changed as expected, but the individual enzyme molecules always exhibited a normal distribution of activities around the mean (Fig. 3.5). As the authors noted (p. 5949), "The large number of single molecule substrate turnover rates is representative of the activity distribution within an entire enzyme population." The supposedly identical molecules in fact consistently exhibited different levels of activity at each substrate concentration, fitting closely to a normal distribution. This is unlikely to be what we would usually call experimental error, which is attributed to the error associated with taking

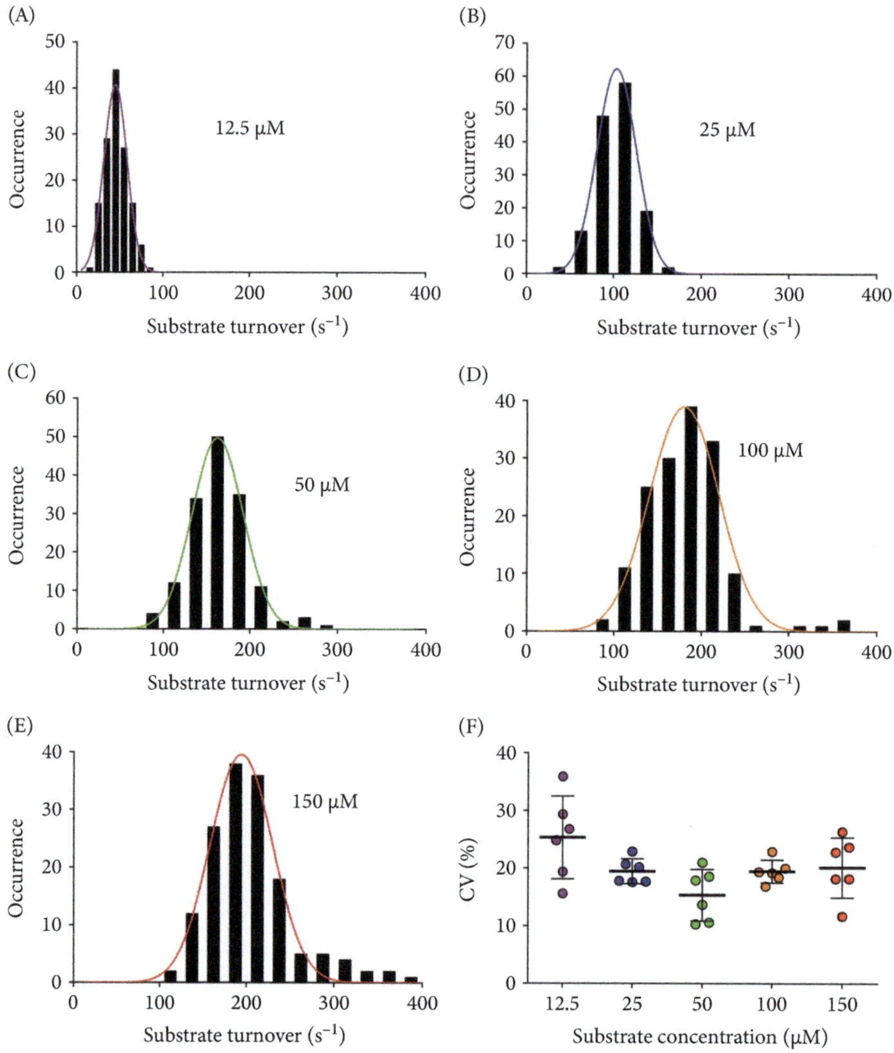

Figure 3.5 Population distributions of enzyme activities (substrate turnover rates) of individual enzyme molecules across a range of substrate concentrations (**A-E**). The activities of single beta-glucuronidase (GUS) enzyme protein molecules were measured individually in femtoliter wells. Even though all proteins were produced in *Escherichia coli* using identical mRNAs, consistent variation among the enzyme population was detected, apparently due to variation in the folding of the amino acid chains into their 3-D structures. **F.** The coefficients of variation (CV, or σ/μ), indicate that the shapes of the normal distributions (colored curves) varied little across substrate concentrations.

Reproduced with permission from Liebherr, R. B., et al. (2014). A single molecule perspective on the functional diversity of *in vitro* evolved β-glucuronidase. *Journal of the American Chemical Society* 136(16): 5949–5955. https://doi.org/10.1021/ja412379p.

multiple small samples from a given large population. Instead, in this case, large numbers of enzyme molecules are being individually assayed, so there is no sampling error per se due to averaging of the individual measurements. In addition, the relative variation among individuals around the mean (coefficient of variation) was constant as the substrate concentrations were changed (Fig. 3.5F), indicating a stable population distribution. The authors attributed this variation to random differences in the folding of the amino acid chains into their final three-dimensional structures, influencing their enzymatic efficiency.

The results of Liebherr et al. (2014) shown in Figure 3.5 are revolutionary for our conception of enzymology, which fundamentally assumes that enzymes with the same amino acid sequence are identical proteins that respond identically to changes in substrate availability and concentration. Enzymes can be activated or deactivated, as by phosphorylation or other post-translational modifications, but these are considered to be state changes, and all phosphorylated versions of a given enzyme molecule would exhibit a common enhanced activity relative to their unphosphorylated versions, for example. The same rationale is also applied to transporters, sensing molecules, and the myriad other functions conducted by proteins in a cell or an organism. While accepting that they can exist in different states, we still assume that individually they are identical as determined by their amino acid sequences. However, the results of Liebherr et al. (2014) indicate that even proteins of identical composition and source can vary in activity due to variations in their structure, apparently resulting from different three-dimensional folding. Thus, the population basis of biology extends to the very structure of presumed identical protein cohorts.

If this postulated variability in protein conformation seems odd, I can recommend reading Natasha Myers's *Rendering Life Molecular* (Myers, 2015), which could be considered as a venture into the philosophy of protein crystallography. We forget that to obtain such structural information, the proteins have to first be induced to crystallize, and the conformations that they assume in this state may not be identical to those they assume in the hydrated, active state. We also know now that protein molecules move and their conformations change as they perform their biochemical functions: "an enzyme molecule is an ever-fluctuating dynamic entity during catalysis" (English et al., 2006, p. 92). In fact, diverse protein structures for the same amino acid sequence (or various modifications of it) can now be predicted by computational models such as AlphaFold3, Evo 2, or BioEmu (Abramson et al., 2024; Brixi et al., 2025; Lewis et al., 2025). It is therefore not so strange to find that populations of proteins with the same amino acid sequences are not necessarily structurally identical under physiological conditions. As the proteins were synthesized in bacteria, it also is possible that they were modified post-translationally, contributing to their variation in activity. In addition, we know now that approximately one-third of proteins encoded in the genome do not have fixed three-dimensional structures, but rather are what are called intrinsically disordered proteins (IDPs), which have been shown to be critical in many cellular processes, including mRNA processing (Weis and Hondele, 2022), stabilizing cell osmotic potential (Watson et al., 2023) and protein concentrations (Klosin et al., 2020; see Fig. 8.2), and enabling tolerance to environmental stresses (Field et al., 2023) through their ability to promote liquid–liquid phase condensates in cells. Thus, we should adjust our expectations about the scope of protein structural variation *in vivo* and view our data with updated underlying assumptions.

The discovery of Liebherr et al. (2014) represents a direct experimental validation of the basic assumption of the PBT model that population-based features underlie many biological phenomena. (Further examples of such inherent variation at the cellular and individual organism levels follow in later chapters.) For the case of enzymes, it implies that when mean values are satisfactory, bulk measurements of populations pooled together may be sufficient, as they do provide the average value across all individuals. However, when we penetrate into the deeper level of the individuals making up those populations, we discover that variation is present. Just as the average value of many throws of two dice will equal seven, that mean value does not encompass all of the information contained in the underlying process (Fig. 2.1). For approximating the overall activity profile of a given enzyme, the Michaelis–Menten kinetics approach with extracted or synthesized proteins may be valid. However, what if the enzyme protein population itself represents a sensitivity distribution with respect to the turnover rate for a given substrate concentration, and likely for binding to regulatory factors as well? In general, the field of enzymology has assumed that the proteins being measured are identical, preferring to attribute kinetic variation primarily to diffusion or other physical factors while assuming a constant binding affinity among all individual enzyme molecules, which is determined as the substrate concentration giving the half-maximal rate of substrate turnover. However, with my population-colored glasses on, the enzyme activity versus substrate concentration curves look like they could be population based, and in fact the PBT model can reproduce the same behavior as does the Michaelis–Menten model when the sensitivity distribution is normally distributed on a logarithmic scale (Fig. 3.6). This wide variation among individual enzyme molecules readily accounts for the shape of the Michaelis–Menten model.

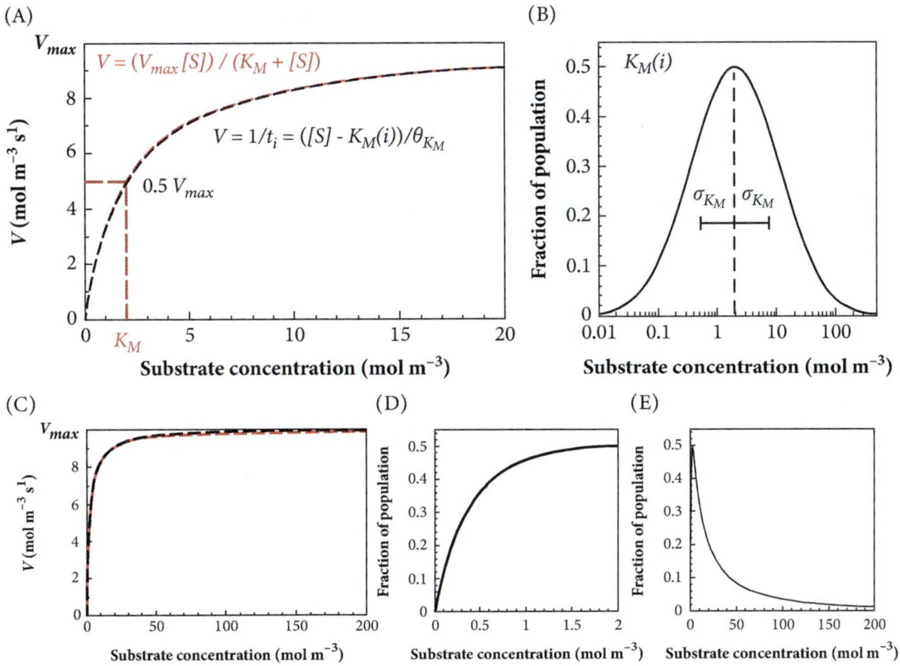

Figure 3.6 Michaelis–Menten kinetics compared to the PBT approach. **A.** The standard Michaelis–Menten model (red equation and red dashed curve) showing an example of the increase in reaction rate [V] with increasing substrate concentration. The Michaelis-Menten constant [K_M] is the substrate concentration at which the rate is half-maximal. The black dashed curve is the reaction velocity [$V = 1/t$] based upon the assumption that K_M varies among the enzyme population [$K_M(i)$] in a logarithmic manner with substrate concentration (black equation). **B.** A normal distribution of enzyme activity thresholds (K_M values for *i* enzyme molecules, or $K_M(i)$) relative to the logarithm of substrate concentration. It has a mean of 2 mol m^{-3} and a standard deviation [σ_{K_M}] of 0.74 log (mol m^{-3}). A time constant [θ_{K_M}] of 0.001 log (moles m^{-3}) s^{-1} was used to generate the PBT curve in panel **A**. **C.** A graph extending panel **A** to high concentrations, indicating that it takes on the order of 50 to 100 times the concentration of the K_M to reach V_{max}, or the maximum velocity in the Michaelis–Menten equation, suggesting that a logarithmic span of enzyme activities is feasible. **D.** Expansion of the $K_M(i)$ distribution up to its peak at the K_M value on a linear substrate scale, representing the median of the population. **E.** Expansion of the $K_M(i)$ logarithmic distribution (panel **B**) across its entire width on a linear substrate scale to illustrate the difference between logarithmic and linear scales for the same sensitivity distribution.

It is also relevant to the occurrence of "enzyme promiscuity" (variation in substrate specificity) that is commonly observed and may be associated with neofunctionalization of enzymes after gene duplication (Khersonsky and

Tawfik, 2010). This underlying catalytic variation also likely contributes to the evolution of the remarkable metabolic variation and chemodiversity of terrestrial plants (Weng et al., 2012) and even to the mating behavior of birds (Loveland et al., 2025). Further examples are shown and discussed in later chapters.

The data from Liebherr et al. (2014) indicate that the variation in activity among individual enzyme molecules is on the order of three-fold (Fig. 3.5). At any substrate concentration, individual enzyme molecules could vary up to three-fold in their activities, even with identical (in this case) amino acid sequences. This provides an opportunity for both the recruitment and kinetic components of the PBT model to come into play (Fig. 2.4), depending upon the environment in which a given molecule is present. Recently, a diffusion-based microfluidic method that enables estimation of binding affinities of individual molecules (DNAs or proteins) to other ligand molecules found variation of 100-fold in ligand binding affinities and rate constants (Zhu et al., 2025). In addition, while estimates vary, an expressed protein in a cell potentially may be present in up to 100 "proteoforms" when nucleotide polymorphisms, post-translational modifications, phosphorylation, pH and so on, are included (Aebersold et al., 2018). Proteomic methods allowing dose- and time-related resolution of protein activity (i.e., as affected by drugs, protein interactors, proteasome and kinase inhibitors, antibodies, etc.) have revealed that variation in post-translational modifications of proteins exhibit population-based patterns influencing their activity, resulting in reaction rate variation over multiple orders of magnitude (Zecha et al., 2023). Multidomain proteins connected by flexible linkers further expand the potential structural heterogeneity inherent in such proteins (Schneider et al., 2023). Furthermore, unlike in a test tube with uniform mixing, individual proteins can be segregated *in vivo* into specific cellular compartments in which their local environments may differ, particularly at the nanometer scale at which diffusion processes dominate on a molecular timescale (Parthasarathy, 2022). The discovery that relatively minor differences in protein structures can influence their ability to segregate into different fluid–fluid phases (condensates) within the cytoplasm is also likely to be relevant here (Klosin et al., 2020; Emenecker et al., 2021). At a minimum, we can ask whether the features (i.e., shapes of dose-response curves) of enzyme activities, as assumed in basic analytic procedures such as Michaelis–Menten kinetics, might need to be reexamined in light of the discovery that consistent and measurable differences exist among individual enzyme molecules, enabling the possibility of population-based behaviors that have seldom been recognized, much less explored, in biochemistry (an exception is English et al., 2006).

An illustrative example of the application of the PBT model to enzyme kinetics is shown in Figure 3.7. A fluorescent peptide reporter enables the monitoring of the activity of a specific protease (PCaspase) in real time

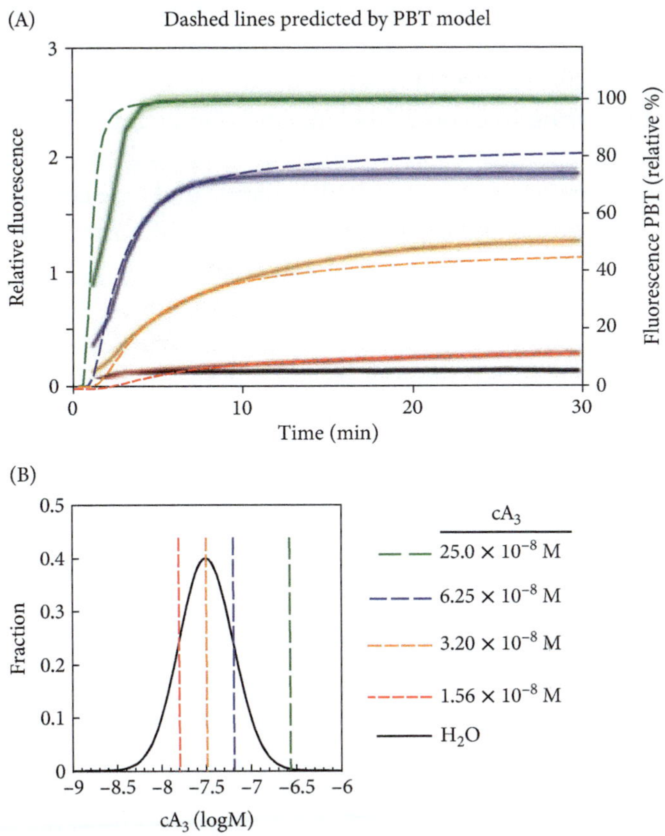

Figure 3.7 Dependence of a PCaspase enzyme on the concentration of its activator molecule (cA_3). **A.** The solid lines are the experimental data, and the shaded regions around them are the standard errors of three technical replicates (from Steens et al., 2024). The dashed lines of the corresponding color are the predicted time courses estimated by empirically fitting the PBT model to the data. **B.** The underlying sensitivity distribution for the predicted curves in **A** based on the concentrations of activator used and the PBT model (mean = −7.5 log M, standard deviation = 0.3 log M; time constant = 1 log M min) fit empirically to the data by the author.

Data in panel A reproduced with permission from Steens, J.A., et al., (2024). Type III-B CRISPR-Cas cascade of proteolytic cleavages. *Science* 383(6682): 512–519. https:/doi.org/10.1126/science.adk0378.

(Steens et al., 2024). Addition of an activator molecule (cA3) increases the rate of change in fluorescence as well as the maximum fluorescence attained (Fig. 3.7A). The shading around the colored lines at each cA3 concentration indicates the standard error of the mean of three technical replicates for each concentration. It is evident that this type of error is small relative to the differences due to the increasing activator concentrations. The pattern of increasing fluorescence is strikingly like the pattern shown in Figure 2.4B based on the PBT model, illustrating the decrease in response time and the increase in maximum response as the activation factor exceeds the response thresholds of an increasing fraction of the enzyme population. This is illustrated in Figure 3.7A by the dashed lines, which are empirical fits to a PBT model using single values for a mean, standard deviation and time constant based on a logarithmic response to the activator. The predicted normal distribution of thresholds in the PCaspase enzyme population is shown in Figure 3.7B, along with vertical lines illustrating the experimental concentrations of the activator (see Fig. 2.4A for comparison). The comment on Figure 3.7A in the original paper noted only that "Sensitivity for cA3 of the FAM-peptide visualization method is 15.6 nM" (Steens et al., 2024, p. 515), the lowest concentration that gave a detectable response. However, calling this the "sensitivity" of this assay ignores that at each cA3 concentration, the fluorescence reaches a maximum plateau that does not increase over time at that concentration. This would not be expected from our usual Michaelis–Menten conception, in which catalytic rates are assumed to be largely determined by diffusion and the effects of substrate concentration on it. However, if all enzyme molecules are equally active at each concentration, why would these curves plateau rather than continue to increase over time until the substrate was consumed? Each concentration would be associated with a different rate of increase, but given enough time, all of them should eventually reach the same maximum level, depending upon the amount of substrate provided. The recruitment aspect of the PBT model accounts for this, as additional active enzyme molecules would be recruited at each increasing activator concentration, and the maximum response would be proportional to the fraction of the enzyme protein population that has been activated (Fig. 3.7B). As it becomes increasingly possible to use such methods to monitor enzyme activities at the molecular level, I expect that more patterns such as these will be reported. I also note that the same type of behavior is observed in inducing or inhibiting germination of seeds by promotive or repressive hormones (see Fig. 5.8). In this case, there is no doubt that the patterns represent the collective behavior of individuals with differing sensitivity, as the seeds can be clearly identified as individuals, each with only one off/on transition in response to the hormonal effector.

Another aspect of this example also bears mentioning. In addition to the minimum sensitivity detected in an assay, enzyme activities are also often characterized by the substrate concentration at which activity is half-maximal (i.e., K_M in the Michaelis–Menten equation; Fig. 3.6). However, in this PCaspase example, the overall response is normal on a logarithmic scale of activator dosage. If the logarithmic distribution of sensitivity dependence is not recognized, then what constitutes "half-maximal" is more complicated (see Fig. 3.6D, E). In this case, the mean activator concentration is −7.5 log M (31.6 nM) and the standard deviation is 0.3 log M. If we double that concentration, we get 63.2 nM, or −7.2 log M, an increase of one standard deviation from the mean, or only 84%, not 100% of the population (see Fig. 2.2). In addition, the logarithmic dependence on concentration results in an asymmetry regarding the variation. In this case, going from 16% (one standard deviation below the mean) of maximum response to 50% would require an increase of only 15.8 nM, whereas to go from 50% to 84% would require an increase of 31.6 nM, twice as much. That is, logarithmic scales indicate that it takes larger changes in concentration to achieve a given increase in activity as the concentration increases; that is, the enzyme molecules requiring high concentrations are increasingly harder to activate (Fig. 3.6). Appreciating the characteristics of an underlying log-normal distribution of sensitivities and understanding its relationship to the concentration dependence clarify what is actually occurring at the molecular level.

Similar considerations apply in pharmacological studies that focus on drug interactions with a receptor molecule. These dose-response relationships have been thoroughly explored for decades and have been quantified mathematically (Neubig et al., 2003). However, I note that among the basic assumptions underlying these universally accepted concepts and equations is one stating that "the response observed is mediated by a uniform population of receptors" (Neubig et al., 2003, p. 601). That is, a fundamental assumption of pharmacological research is that the receptors under study with respect to a given binding factor are all identical. Hormonal action in endocrinology also makes this assumption, although there are abundant data on interindividual variation in hormonal levels and responsiveness (Williams, 2008). In contrast, an underlying principle of the PBT approach is that the individuals under study are not identical but rather vary for some of the features under study, often (but not necessarily) in a normal distribution (on a linear or log scale). This assumption of uniformity at the molecular level is analogous to some of the fundamental assumptions underlying physics, such as the uniformity of fundamental particles. Such assumptions work well at certain levels of examination, but, as with the distinctions between Newtonian and quantum physics (McFadden and Al-Khalili, 2014), the actual situation is more complex when considered under other conditions (Yvert, 2014; Kauffman and Roli, 2023). Nonetheless, the mathematical conventions dependent upon those simplifying assumptions can take on a reality for scientists that overshadows what is actually being observed, a situation that has been termed "The Blind Spot" by Frank et al. (2024). As these authors argue, once certain mathematical concepts, such as thermodynamic principles (or Michaelis–Menten kinetics), have been firmly established based on associated assumptions, those concepts take on a reality of their own, and the underlying assumptions on which they are based are no longer recognized or even seen.

An illustration of this "Blind Spot" with respect to protein receptors and their agonists (chemicals that trigger a positive response in the receptors) involves G protein (heterotrimeric guanine nucleotide-binding protein)-coupled receptors, or GPCRs. Such GCPRs "constitute a major family of membrane proteins that respond to diverse extracellular ligands, including photons, small molecules, neurotransmitters, and hormones. In humans, more than 500 endogenous ligands act on 800 GPCRs and regulate key aspects of human biology" (Heydenreich et al., 2023, p.1). In a remarkable study, Heydenreich and colleagues systematically modified each of the 412 amino acids in a single β2 adrenergic (responsive to adrenalin) receptor–G protein (β2AR) and studied the key pharmacological properties of each mutant protein. The pharmacological properties assessed were *potency* (concentration producing half-maximal response) and *efficacy* (maximum response attained) as determined by concentration-response curves on a logarithmic scale (Fig. 3.8), which are common in pharmacological studies (e.g., Tan et al., 2024; see also Fig. 6.7). The potency parameter is the mean of the normal distribution in sensitivity to adrenalin, which is equivalent to the 50% point in the sigmoid cumulative curve. The shift to the right (without changing the shape of the curve) indicates that the mean of the sensitivity distribution has shifted to a higher concentration, but the variation among the individual sensor β2AR proteins has not changed (i.e., the standard deviation is the same). In this interpretation, potency is equivalent to the mean of the threshold distribution of binding of adrenalin to individual β2AR proteins, which in this interpretation are not identical. As shown above for enzyme activities (Fig. 3.5), the fact that the proteins in a given test have identical amino acid sequences does not mean that they must therefore have identical binding sensitivities, particularly under *in vivo* conditions. The amino acid sequence of a protein does not fully determine its folding and 3-D structure, nor does it account for any post-translational modifications. Similarly, the maximum activities (efficacies) of the β2AR mutant proteins varied (Fig. 3.8B), which likely indicates that some conformations bound adrenalin, while other conformations did not or did so with less binding force or stringency. Different maximal values at high agonist concentrations likely indicate the inability to activate all of the protein structural variants present under the test conditions. Thus, both the potency and the efficacy determinations appear to be based upon variation in the properties of the proteins themselves; that is, they are not all identical, which is the underlying hypothesis of the PBT approach. The ability of the PBT model to deal directly with shifts in both mean and maximum rates as being dependent upon the underlying sensitivity variation among the individual proteins in the population (e.g., Figs. 3.6, 3.7) provides strong evidence that it should be considered as an alternative to the "uniform population of receptors" assumption that currently underlies pharmacological studies. This does not invalidate the information learned from analyzing only means (e.g., potency), but the results are incomplete when considering the actual situation at the functional level. The inability of biologists to see the population variation that is evident in their data (over orders of magnitude as indicated by log scales) constitutes a Blind Spot (Frank et al., 2024) that I will return to repeatedly in this book.

I also note that if we extend this underlying concept to include 800 GPCR family members, each of which could exhibit a unique and variable sensitivity distribution to 500 endogenous ligands involved in a range of essential biological processes (Heydenreich et al., 2023), the potential complexity of biological reality becomes evident. However, my contention is that organization into populations exhibiting a sensitivity distribution for a given ligand brings order, rather than disorder, to the situation. Stochasticity exists at every level in biology, yet biology could not function with unbridled randomness. Rather, biology adapted to the inherent stochasticity at its fundamental levels by utilizing the uncontrollable variation as a feature, not a bug. The reservoir and recruitment components of having diverse individuals available offers manifold possibilities for overlapping responses and regulation, while the kinetic aspect of more rapid responses as the effector level increases above individual thresholds enables coordinated timing of biological processes in response to sensed conditions.

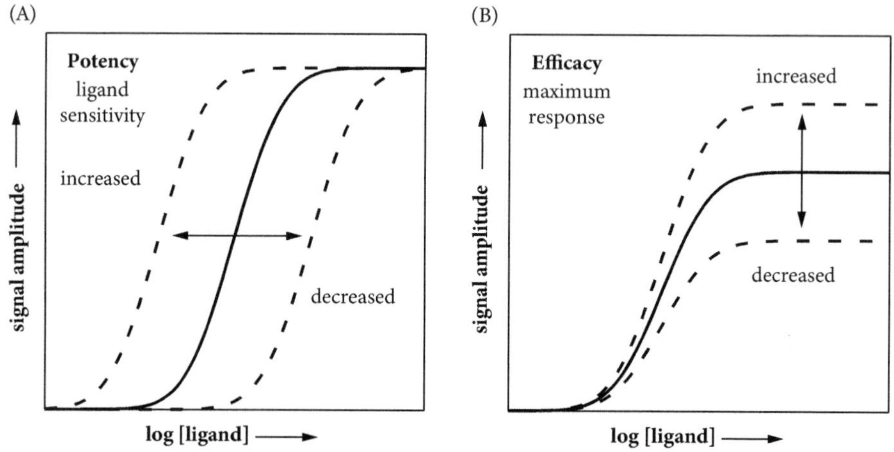

Figure 3.8 Dose-response curves (solid lines) for ligands and their binding sites illustrating changes in either potency (**A**) or efficacy (**B**) in triggering a response. **A.** Changes in potency are determined by shifts in response curves to lower or higher values (dashed curves). **B.** Changes in efficacy are indicated by the maximum activation response to the ligand. Note that in both panels, the concentration response exhibits a sigmoid curve, or the cumulative version of a normal distribution, on a log [ligand concentration] scale.

Adapted with permission from Heydenreich, F.M., et al. (2023). Molecular determinants of ligand efficacy and potency in GPCR signaling. *Science* 382(6677): eadh1859. https:/doi.org/10.1126/science.adh1859

Conclusions

Our current understanding of the origin and integration of biological processes is still at an elementary stage, but even so, it represents a remarkable achievement of human consciousness, observation, and inquiry to reach its present state. Reasonable and rational hypotheses exist for how organic molecules could be formed prior to biology, how these could form "collective autocatalytic sets" that could self-reproduce, and how biology could evolve into the current biosphere (Kauffman, 2019; Jablonka and Lamb, 2020). Initially, randomness and stochasticity would have ruled, but it would have been too slow to try all possibilities and would have been a counterforce to reproduction. Thus, early biology would have needed to deal with stochasticity without being able to do much to control it. The PBT approach accepts this and includes such random variation as a feature, not a bug in the system. In fact, the occurrence of population distributions exhibiting systemic variation in key properties creates a standing "adjacent possible" for further exploitation by biological systems (Kauffman, 1993, 2019). The earliest biological organisms of which we are aware already exhibit key aspects of PBT function and behavior. And evolution is well known to preserve things that work and progressively modify them, rather than starting over again with an entirely novel approach. Thus, as we move through the evolutionary hierarchy of biology in the following chapters, we will repeatedly encounter mechanisms and behaviors that can be understood as manifestations of the underlying extant variation and how stable reproducing systems have co-opted it to a wide diversity of uses.

4
Microorganisms

Cellular individuality in populations

The jump from the molecular world to the cellular world was momentous and profound, and it is still unclear exactly how it was accomplished (Szostak, 2017). Nonetheless, it occurred multiple times, and at least two major lineages persist today from that period, nearly 4 billion years ago, the prokaryotes and the archaea (Gribaldo and Brochier-Armanet, 2006). Both consist of single-celled organisms lacking nuclei, the defining cellular organelle of the third major lineage of biology, the eukaryotes (fungi, plants and animals). The key biochemical pathways essential for life are present already in the prokaryotes and archaea (I will not distinguish between these here), as well as the ability to reproduce by cell division (Ku et al., 2015; Zaremba-Niedzwiedzka et al., 2017). In addition, the evolution of cells fundamentally changed the underlying biological processes by isolating them from external conditions due to the development of membranes that enabled some control over internal conditions (solutes, pH, etc.). However, in both single-celled and multicellular organisms, the cells are not all identical.

Thus, a major question in biology is how do all of these cells in a given species, each of which has the same (or very similar) genome, nonetheless accomplish all of the diverse tasks required to live and multiply (at the single-cell level) or to diverge into all of the different cell types required for a functional multicellular organism? Alfonso Martinez Arias has explored this issue in his book *The Master Builder* (Arias, 2023) and makes the argument that cells are the core of biology, not DNA or genes per se. He views the genome as basically a toolbox encoding proteins that are accessed as needed within cells to perform specific functions, construct diverse cell types or adapt to the current environment, not as the dictator calling all the shots to subordinate processes or cells. In fact, it seems from the previous chapter that DNA more closely resembles a hereditary king that is essentially captive in his palace, constantly served by a multitude of attendants that manage all aspects of his life and carry out all of the bureaucratic functions derived from his powers. A similar analogy for DNA would be a piano, which has a fixed number of

Populations of Individuals. Kent J. Bradford, Oxford University Press. © Kent J. Bradford (2026).
DOI: 10.1093/9780197846032.003.0004

strings that can each play only one note (comparable to the individual genes in the genome), but it is the pianist (the cell) that determines which notes are played at a specific time, for how long, at what volume, and so on. And yet, think of the diversity of music that can be created on a piano that has only 88 keys, only 10 of which can be played simultaneously by a single pianist. And then compare that musical variation possibility to the potential of the tens of thousands of genes in a genome, half of which are being expressed at any given time. This dynamic relationship between individuals and their associated collective interactions provides a rich basis for "creative evolution" as described by Henri Bergson in 1907 even as the concept of genes was just beginning to be appreciated (Bergson, 2016, p. 284): "Thus, in rudimentary organisms consisting of a single cell, we already find that the apparent individuality of the whole is the composition of an undefined number of potential individualities potentially associated. . . . The evolution of life in the double direction of individuality and association has therefore nothing accidental about it: it is due to the very nature of life."

In this chapter, I will begin after the evolution of individual cells and look at population aspects of their ability to acquire nutrients, grow, adapt, multiply and collaborate. In general, microorganisms have been studied in mass, that is, by culturing them in large quantities and sampling the resulting populations. In many cases, such cultures are sampled in "log phase," or periods in which cell division is occurring rapidly. Since the cycles of cell division are not synchronized in these cultures, which can occur as often as every 20 minutes in bacteria, any bulk sample taken from log-phase cultures will contain cells at all stages of the cell cycle. This is fine if you only want to know what can happen at any stage of cell development, but it can be misleading if you do not remember that it is a highly variable, mixed sample. As noted by Kreft et al. (2013, p. 18027), "Possibly the most important finding for the future of the discipline of microbiology has been the realization that genetically identical bacterial cells in a well-mixed environment may have individually differing phenotypes. This calls into question the validity of population-level experiments and population models that assume that all individuals are the same." In fact, there is an entire field of study called "microbial individual-based ecology" that "is grounded in the notion that to truly understand the interactions of microorganisms with their biotic and abiotic environment, one cannot ignore the scales at which such interactions occur" (Leveau et al., 2018, p. 1). Thus, I will look at a few cases in microorganisms in which individual cell behavior underlies the observed population-level behavior and see how the PBT concept can be applied to them.

Chemotaxis

Two key concepts in the PBT model are that sensitivity thresholds to various signals exist and that these thresholds can vary among individuals. This is illustrated in a classic behavior of single-celled bacteria called chemotaxis: the ability to swim up a gradient toward higher concentrations of a desired chemical (attractant) or away from higher concentrations of an undesired one (repellent). Simple experiments can be conducted by exposing bacterial cultures to different concentrations of an attractant in their growth solution compared to the concentration in an accessible capillary and quantifying how many additional cells accumulate in the capillary (Mesibov et al., 1973). These experiments showed that the response range laid between a threshold concentration and a saturating concentration (Fig. 4.1A), as in the PBT model. This response range varied for different attractants and their corresponding chemoreceptors, indicating varying thresholds and response ranges for different attractants or their receptor proteins. The shapes of these curves generally approximated normal distributions when the attractant concentration was plotted on a logarithmic scale (Fig. 4.1B).

The results shown in Figure 4.1 were interpreted "in terms of the law of mass action as applied to interaction between a chemoreceptor and its substrate"

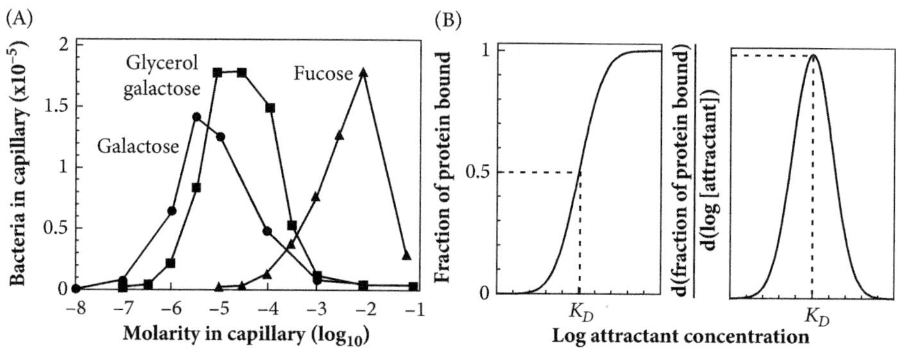

Figure 4.1 Chemotaxis, or the ability to move toward higher concentrations of an attractant, in bacteria. **A.** The number of bacteria that entered a capillary containing a given concentration (molarity) of an attractant in excess of the number entering in the adjacent lower concentration of the attractant, plotted on a logarithmic concentration scale. **B.** The left panel illustrates the theory that a cumulative sigmoid curve of chemosensor proteins bound to the attractant, based on a log-normal distribution (right panel), underlies this behavior.

Adapted with permission from Mesibov, R., et al. (1973). The range of attractant concentrations for bacterial chemotaxis and the threshold and size of response over this range. *Journal of General Physiology* 62(2): 203–223. https:/doi.org/10.1085/jgp.62.2.203

(Mesibov et al., 1973, p. 214), as is generally the case. The functional interpretation (Mesibov et al., 1973, p. 216) was that "the number of chemotactic events is proportional to the difference in fraction of binding protein complexed at the two concentrations, and the number of bacteria accumulating in the capillary is proportional to the number of chemotactic events." That is, the variation with respect to concentration is due to an increasing fraction of a fixed number of receptors being bound to the attractant molecule. This results in the definition of the dissociation constant (K_D) as the concentration at which half of the receptors are occupied (Fig. 4.1). However, one must ask why some of these receptors can respond to 10^{-7} M of galactose, for example (Fig. 4.1A), while a similar fraction remains unbound at 10^{-3} M. Why would some receptors (and the cells containing them) respond at such low concentrations while others required a 10,000-fold higher concentration to respond, assuming that all the receptors are identical and all the cells contain an identical number of receptor molecules, estimated by the authors to be 50,000 per cell? The authors further estimated that about 6000 receptors per cell would need to be occupied by the attractant to produce a detectable response. Since the response is symmetrical (on a log scale), this means that the cells with the most sensitive 6000 receptors would respond between 0.1 and 1 μM, while the cells with the least sensitive 6000 receptors would respond between 100 and 1000 μM. What is the basis of this sensitivity difference if all the receptors are identical and all the cells have equal numbers of them?

It turns out that the bacterial two-component signaling receptor system underlying chemotaxis is considerably more complex (Falke et al., 1997). It is based on a population of protein sensors in the cell that can bind to the target, or attractant; presumably, the fraction of sensors bound increases in a sigmoid pattern as the attractant concentration increases logarithmically. Many physiological responses, such as enzyme activation and hormonal regulation, also exhibit this sigmoid dependence upon the logarithm of the factor concentration (e.g., Figs. 3.7, 3.8, 5.7, 6.2). But this is only the initial step in the process. Having detected the presence of the chemical, how does an individual bacterial cell move in the direction in which the attractant concentration is increasing? Bacteria such as *Escherichia coli* are very small (2 μm long by 1 μm in diameter) (Karmakar, 2021), so the difference in concentration of a chemical in solution from one end of a single cell to the other is far too small to be measured directly. That is, a bacterium cannot just compare the signal from a receptor at one end of the cell to the signal from the other end and move in the direction with the higher value. In addition, although they have flagella and can swim, they must work hard to make headway in relation to Brownian motion and diffusion in solutions. First, they must increase the signal gradient in some way to detect whether it is increasing or decreasing. So what do they do? They take a measurement of the attractant concentration, then swim in a random direction for a period of time (called a "run") and finally reverse the direction of their flagella and "tumble" in place, measuring the concentration again. If there is sufficient signal to indicate an increasing concentration gradient, the bacteria will swim longer the next time before tumbling. In contrast, when the concentration has decreased, they swim for a shorter time before tumbling. This combination of periodic measurements and longer and shorter runs based on the change in the attractant level allows the bacteria to move progressively up a concentration gradient in a "biased random walk." To move away from a repellent signal, this process is reversed and the bacterium swims longer when the concentration has decreased between samplings (Karmakar, 2021). That this biased random walk system can work is shown by the tens of thousands of bacterial cells attracted into the capillaries, as shown in Figure 4.1A.

But even this is not enough to solve all the problems in chemotaxis. As noted above, the sensitivity varies with the logarithm of the concentration. Thus, as the bacterium, by starts and stops, succeeds in moving to where the concentration is higher, its original sensitivity distribution eventually would be saturated, and no more signal would be perceived at higher concentrations. If a bacterium could detect the attractant at a concentration of 10^{-7} M (Fig. 4.1), its sensors would need to be sensitive to small changes in the vicinity of this concentration in order to influence the runtime signal. As the cell made progress up the gradient, the attractant concentration might rise 10- or 100-fold, saturating those sensors and providing no information about whether the external concentration was increasing or decreasing. Again, the bacteria adapted by being able to shift their threshold sensitivities to adjust to the external concentration, keeping their sensors in a range in which they could detect concentration differences within the distance of a bacterial run (Falke et al., 1997). This is achieved through a relatively simple mechanism of phosphorylation and methylation of the receptors that is dependent upon the concentration of the attractant or repellent chemicals. Efficiently, this response system for setting the sensitivity threshold also constitutes a memory system preserving the last concentration measurement, another essential component of the chemotaxis process. This becomes even more complicated when we realize that many receptor proteins for specific chemicals are present at the same time, each representing a population of sensors with their own sensitivities, providing input into the protein motors driving the flagella and determining whether they should be in run or tumble mode at a given point. Even so, "It should be noted that the [adaptation] enzymes act globally on all the chemotaxis receptors in the cell. Thus, when fully adapted, the net methylation level of the receptor population exactly balances the signals generated by multiple attractants and repellents" (Falke et al., 1997, p. 465).

Although such balancing may occur within a given cell exposed to a specific effector concentration, individual cells still exhibit distinct response patterns. For example, "There is great variability in the time that individual bacteria take to return to their prestimulus tumbling pattern. . . . [The] form of the recovery of a population from a stimulus is caused by the variability of the individuals in the population" (Spudich and Koshland, 1976, p. 468). These authors further note (p. 470) that "[s]tudy of individual bacteria in various ways has established that a bacterial culture which is genetically homogeneous and grown in homogeneous nutrient conditions nevertheless produces characteristically different individuals. These individuals retain their properties throughout their life cycle, showing that the individuality is not a momentary effect of random chance bombardments by a fluctuating environment." In fact, Koshland (1980) wrote an insightful review exploring the fundamental relationships between bacterial chemotaxis and the functions of individual animal neurons. However, as discussed by Eckert et al. (2024), the subsequent "lack of impact" and "invisibility in the literature" of Koshland's work were due in part to the prejudicial assumption that adaptation by learning could occur only in organisms having a nervous system. In connection with our theme here, it is worth quoting a key section of Koshland's 1980 paper (pp. 67, 70; italics added for emphasis):

> The existence of such variability leads one to ask whether it has some advantage to the organism or whether it is a disadvantage that must be surmounted in the development of the species. Although there may be no significant selective pressure against nongenetic variability, one can certainly make an argument that nongenetic variability aids in the survival of a population subjected to widely varying conditions during its lifetime. If there were individual bacterial variation in a genetically homogeneous population, a few of the bacteria might either be supersensitive or insensitive. These bacteria would normally not survive as effectively as the bulk of the population, in most circumstances an unimportant loss if the main body of the colony survived. If, on the other hand, unexpected toxic and lethal situations arose, a small fraction, *on the wings of the probability distribution*, might constitute the only bacteria to survive and would then reproduce the species for future generations. Thus nongenetic variability would be a preferred mechanism for accommodation to random fluctuations in the environment and genetic variability [would be] the preferred mechanism for accommodation to long-lasting environmental changes. Induction and repression of enzyme synthesis

> reflect intermediate accommodations to fluctuations of a lengthy but impermanent type. . . . In the chemotactic system, *heredity, environment, and chance each play a significant role*. Certain parts of the system are *inherited* and are the same in every bacterium unless some mutational deficiency appears. Certain other components of the system, particularly certain receptors, are *induced by the growth conditions* of the cell. Finally, *there is an element of chance in the development of the system, which leads to different behavioral responses*.

Here, Koshland has presaged a key aspect of the PBT concept: that genetics and environment interact with underlying developmental variation that creates sensitivity distributions among individuals in a population. All three aspects are evident already in the chemotactic capacity of single-celled organisms.

Koshland's vision is further supported by experiments with growing fungal hyphae of *Aspergillis nidulans* (Yamamoto et al., 2024). Using a microfluidic system in which the growing hyphae (single cells) could choose to grow in distinct solutions on one side or the other of the device, these authors demonstrated chemotropic behavior by the hyphae, which avoided high concentrations of nitrate and ammonia. This selection required the presence of membrane transporters for these nitrogen compounds, which occurred throughout the cell, but it was controlled specifically at the hyphal tip, where extension growth occurs. Thus, consistent with the PBT concept, the wide variation observed in dose-response curves is more likely due to sensitivity variation among (or within) the individual cells making up the population, rather than being explained by biophysical mass action models based on identical biological entities, whether receptor molecules or cells. An example of the latter approach is the work of Eckert et al. (2024) modeling habituation in single-cell learning. Habituation occurs when the response to a repeated stimulus steadily declines with repetition, and it is considered to be one of the simplest forms of learning (Lyon, 2015; Dussutour, 2021; Lyon et al., 2021). Eckert and colleagues developed and demonstrated various interactive networks of the biochemical intermediates between a stimulus and response that can reproduce the habituation response to repeated stimulation. However, as with virtually all biochemical models, the underlying entities in these networks are assumed to be identical, and their differences are characterized by, for example, Michaelis–Menten kinetics and K_M values (see Fig. 3.6). This assumption does work, although it is rather complicated, but it does not account for the emphasis by Koshland (1980) on the importance of individuality and variation in his studies of single-cell population behavior. This basic chemotactic system has also been analyzed using the agent-based modeling approach described in Chapter 2, which does explicitly incorporate individuality among the bacteria (Korobkova et al., 2004; Emonet et al., 2005). However, when it is considered, individual variation is generally introduced as a stochastic randomness generator in a model just to provide the expected variation or "error," rather than as a feature of a structured population of varying individuals constituting a sensitivity distribution.

Additional work on the habituation response in a single-celled ciliate (*Stentor coeruleus*) illustrates how observations of individual cells can contribute to this issue. This unicellular organism responds to mechanical stimulation by contracting into a ball. A high percentage of cells responds in this way to an initial stimulus, but the fraction of responding cells decreases with repeated episodes, indicating habituation to the stimulus (Wood, 1969). By monitoring individual cells, Rajan et al. (2023) demonstrated that the overall behavior of the population was determined on a quantal basis. That is, individual cells differed in the number of repeated stimuli required to trigger their contraction, but once triggered, those cells rarely returned to their initial sensitivity. That is, the response was quantal (cell-by-cell switching) rather than graded (each cell gradually becoming habituated). With much longer intervals between the mechanical stimulation, the population could return to its initial sensitivity. It is evident that this is exactly the behavior that would be consistent with the PBT model in which cells would vary in their sensitivity to habituation and increasing stimulus events would result in increasing numbers of cells becoming habituated (i.e., ignoring the signals). As the authors noted (p. 244), "We conclude that although a population of *Stentor* cells shows gradual habituation, the response of individual cells may be better described by a single switch-like transition from a responsive to a non-responsive state, rather than a graded response." However, their analyses do not utilize a population-based approach, except to the extent that stochastic models are compared to the individual switch-like model to support their conclusions. I have therefore used the PBT approach to model their data (Fig. 4.2). My assumption is that the individual cells vary in a log-normal distribution in their sensitivity to the stimulus events (see, e.g., Figs. 3.2, 3.6). This simple assumption fits the habituation data well, shown as the declining fraction of responding cells as originally presented (Fig. 4.2A, C), or as an increasing fraction of habituated cells (Fig. 4.2B, D), based simply on the cumulative sensitivity distribution of individual cells in the populations. Changes in the values of the mean and standard deviation of the sensitivity distribution were sufficient to account for differences in response due to the magnitude of the stimulus (high force versus low force). While not in itself contributing to

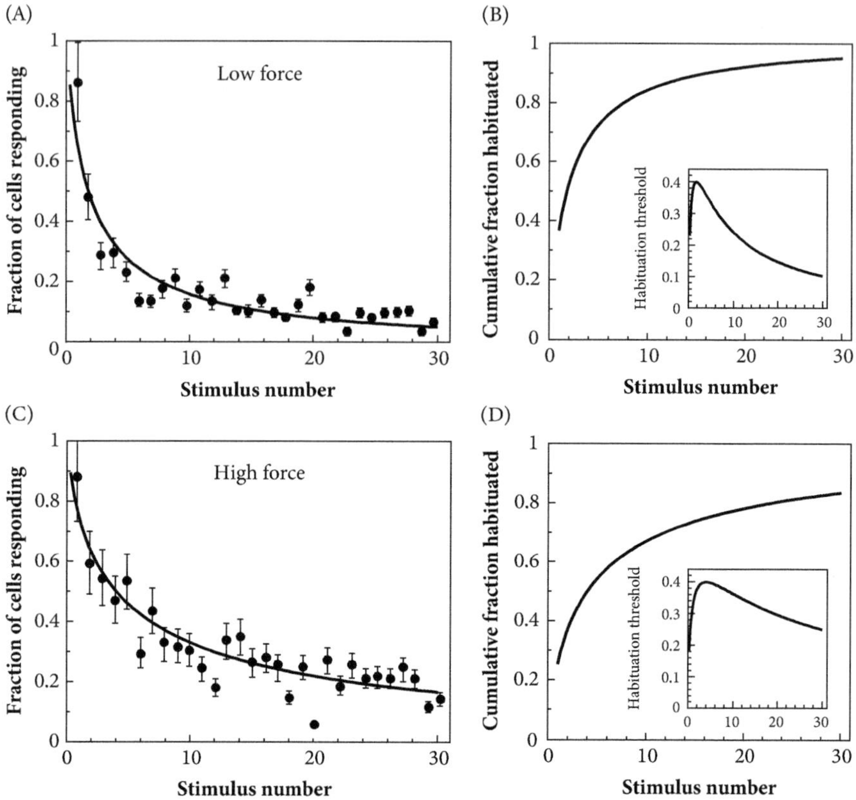

Figure 4.2 Habituation of *Stentor coeruleus* cells to mechanical stimulation. The cells of this ciliate contract into balls upon mechanical stimulation. With repeated stimulation (once each minute), the fraction of cells responding declines, indicating that they have become habituated and ignore the stimulus. **A.** Data for responses to low mechanical force; the solid line is predicted based on panel **B**. **B.** Cumulative fraction of habituated cells as stimulus number increases, based on the sensitivity distribution shown in the inset graph (a log-normal distribution plotted on a linear scale; mean = 1.8, standard deviation = log [5.5]). **C.** Data for responses to high mechanical force; the solid line is predicted based on the sensitivity distribution shown in **D**. **D.** Cumulative fraction of habituated cells as stimulus number increases, based on the sensitivity distribution shown in the inset graph (a log-normal distribution plotted on a linear scale; mean = 4, standard deviation = log [8]).

The data in panels A and C are reproduced with permission (Open Access) from Rajan, D., et al. (2023). Single-cell analysis of habituation in *Stentor coeruleus*. *Current Biology* 33(2): 241–251. https://doi.org/10.1016/j.cub.2022.11.010.

understanding the mechanisms underlying such habituation responses, it is evident that the overall behavior of habituation is readily described by the PBT approach, that is, sensitivity distributions triggering a quantal change in individuals, indicating that looking for such population-based mechanisms could be fruitful, as suggested by Koshland (1980).

Quorum sensing

In some of the fundamental behaviors of single-celled organisms, chemotaxis and habituation, we can see both the threshold-based and population-based features of the sensitivity distributions governing them as well as the ability to adjust those thresholds in response to information coming from the environment. This indicates the remarkably high level of biological sophistication already present in molecular signaling systems in single-celled organisms. The extension of this signaling system beyond single cells is also evident in the phenomenon of "quorum sensing," or the ability for sensitive individual cells to signal to others that something needs to happen collectively. Quorum sensing occurs when small molecules secreted into the environment by individual cells reach a certain threshold that triggers a behavioral change or developmental response in other cells in the population. Originally, it was assumed that individual cells were uniform in their secretion of their inducer molecules (i.e., "autoinducers"), which were therefore thought to primarily indicate the number of cells present in a local population, with the subsequent response triggered in all cells once the threshold concentration was exceeded. More recently, Bettenworth et al. (2022, p. 2) noted the following:

> However, over the past decade several cases of cell-to-cell heterogeneity in autoinducer synthase or precursor gene expression have been reported. . . . Both the precise nature of these heterogeneities—whether they represented stable subpopulations with distinct expression levels, or rather variations over time—and their molecular origins remained unclear, but their observation nevertheless indicated that the model of constitutive autoinducer production is not universally valid. . . . . [In fact,] phenotypic heterogeneity in autoinducer synthase gene expression and physiological influences on quorum sensing are linked in a collective decision-making process in which the first represents the key for integration of the latter.

Bettenworth et al. further showed that individual cells vary not only in their release of the autoinducer signals, but also in the "stochastic pulsing" of the production of the autoinducers themselves. However, as the authors note (p. 7), "the term 'noise', albeit widely used as a synonym for stochasticity . . . does not seem appropriate in this context, as it has connotations of mere statistical fluctuations." They further (p. 7) compare the signaling process to "voting in a local community," in which, "due to the common pool of autoinducers—comparable to a ballot box—the pulse frequencies of all members of the population are then integrated into the total [signal] concentration; only if this vote crosses the threshold, the response behaviour is initiated." I will return to this analogy of voting populations later (in Chapter 8), and the authors here have presaged another aspect that I will explore subsequently: that the total population is subdivided into distinct subpopulations, with the behavior of the overall population being the sum of the contributions of those subpopulations (e.g., Fig. 5.12). It is clear that the most recent elaboration of quorum sensing in microorganisms is fully compatible with the tenets of the PBT model and that stochastic processes might be better understood as enabling a population distribution of autoinducer production and pulsing, along with corresponding distributions of receptor sensitivities.

Multicellularity

The existence of multicellular responses by populations of independent cells (i.e., quorum sensing) was known prior to the development of the latter term (Fuqua et al., 1994). The classic example is the ability of the amoeba *Dictyostelium discoideum* (slime mold) to grow as independent cells (amoebae) when resources are plentiful, but to aggregate together and form a multicellular fruiting body that produces spores when food is limited (Fig. 4.3). These spores are disseminated, and they can germinate into individual amoeba, completing the life cycle (Nagano, 2000). The key to this behavior is the ability of cells to sense and census their neighbors and respond appropriately to aggregate together. This intercellular communication is based on synthesis and release of cyclic adenosine 3',5'-monophosphate (cAMP) signals by the individual amoebae, with the extracellular concentration of cAMP indicating the number of other *D. discoideum* cells in the vicinity of a given cell and their directionality from the sensing cell (Parent and Devreotes, 1999). This communication system essentially uses the signal-receptor model of chemotaxis described above, with the modification that the signal is produced and secreted by each cell, so that the external concentration of the signal increases as more cells aggregate into smaller spaces. In addition, the receptors for cAMP turn off for a period of time after binding the signal, and then they turn back on. This intracellular receptor cycling results in synchronization of the synthesis/secretion patterns of the cells, further increasing local cAMP concentrations and promoting their aggregation (Nagano, 2000). Due to these properties, *D. discoideum* has become a key model system for

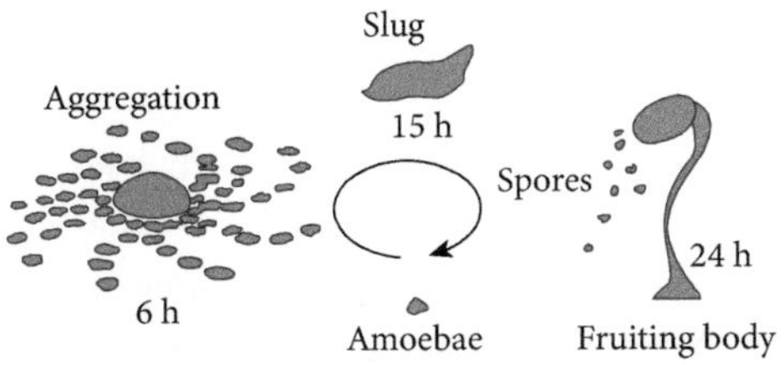

Figure 4.3 Life cycle diagram of the slime mold *Dictyostelium discoideum*. Single-celled amoebae live independently when food resources are abundant. However, when food becomes scarce, the amoebae within a local region aggregate together to form a multicellular slug, which then differentiates within ~24 h into a stalk, with a fruiting body producing spores. The spores (more resistant to unfavorable conditions) are then dispersed and can subsequently germinate to form new amoebae when they sense improved conditions.

Reproduced with permission from Nagano, S. (2000). Modeling the model organism *Dictyostelium discoideum*. *Development Growth & Differentiation* 42(6): 541–550. https://doi.org/10.1046/j.1440-169x.2000.00547.x.

studying chemotaxis, which is involved in a wide variety of physiological and cellular processes throughout eukaryotes and in processes related to various plant and animal pathologies.

Although the key features of chemotaxis in *D. discoideum* have been described by researchers over the past fifty years, the recent development of microfluidic devices that allow the isolation, manipulation and quantification of the movements of individual cells is revealing more precise information about their abilities (Wang et al., 2012; Muljadi et al., 2022). Such discoveries include that there is a threshold concentration of cAMP signal to trigger directed movement, and that above this threshold, the cells moved faster as the cAMP concentration increased. In addition, the threshold concentration varied among different individual cells, with some cells responding to concentrations as low as 10^{-11} M (Wang et al., 2012). These features are clearly embodied in the PBT model. In addition, the authors note the following (pp. 5, 8):

> The model predicted . . . that at low cAMP gradient values, only the [high sensitivity] responder cells might be effectively sensing the cAMP gradient. These predictions were supported by experimental observations that cell polarization and ensuing cell migration up the gradient of cAMP was a direct function of cell responsiveness threshold. . . . Our results suggest that the cell individuality extends further, with the amoebae having a population of more responsive cells that provide the initial "skeleton" of the network of cell responses because they are more sensitive than

> their peers to the input gradient and that transmit the signal from the center(s) of high cell density. This differential sensitivity property can help enhance self-organization of the colony while avoiding potentially inappropriate activation of most of the colony cells by weak extracellular cAMP fluctuations. We thus conclude that variability in the local cell density . . . is not the only determinant of the colony self-organization process, but that this process is also controlled by the fraction of hypersensitive cells that can naturally serve to enhance and nucleate the initial stages of colony response. It will be of interest to explore whether similar phenomena are also present in other eukaryotic cell systems displaying chemotactic behaviors by means of homologous signaling processes.

The implications of this differentiation of sensitivity subpopulations also are directly relevant to the PBT approach, as is evident in the comments of Tsimring (2014, p. 8) on the Wang et al. results:

> Until recently, most of the data on signaling response was collected using population-averaging methods such as Western blots or microarrays, but these approaches can often mask the important dynamical and stochastic aspects of the system response. For example, graded average response to an increased concentration of a signaling molecule may be an indication of a graded response of individual cells or a binary, but heterogeneous response with increased probability of the on state at higher concentration of the inducer [see Fig. 4.4]. A graded chemotactic response to a transient cAMP stimulation is usually observed in a population of amoebae *Dictyostelium discoideum*. However, recent single-cell microfluidic experiments (Wang et al, 2012) showed that the graded population-averaged response is, in fact, caused by a continuous increase of the fraction of on cells in an essentially bimodal population of cells.

The existence of a more sensitive subpopulation of cells to first detect changes in the environment and trigger adaptive responses in the larger population makes clear sense, particularly when considering that distinct hypersensitive subpopulations might themselves be adapted to monitor different environmental signals. In addition, the presence of positive feedback responses combined with high sensitivity to an inducer can result in all-or-none downstream responses once the sensitivity threshold is exceeded in an individual (Ferrell and Machleder, 1998). Such switch-like behavior of mRNA production among different cells has been demonstrated in plants (Alamos et al., 2021).

However, the population-based approach of Wang and colleagues is not applied universally. For example, Ohtsuka et al. (2021, p. 131) noted that

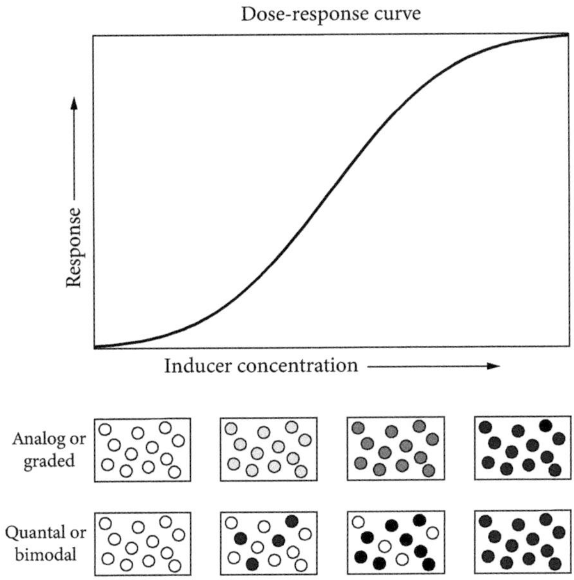

Figure 4.4 Dose-response curves can be produced by two alternative possibilities of signaling responses in a population: a uniform analog or graded response of all cells or a quantal or bimodal but heterogeneous response among individual cells. Inducer concentration is often on a logarithmic scale.

Adapted with permission from Ferrell, J.E., and Machleder, E.M. (1998). The biochemical basis of an all-or-none cell fate switch in *Xenopus oocytes*. *Science* 280(5365): 895–898. https://doi.org/10.1126/science.280.5365.895 and Tsimring, L.S. (2014). Noise in biology. *Reports on Progress in Physics* 77(2): 026601. https://doi.org/10.1088/0034-4885/77/2/026601.

"systematic observations of a large number of individual cells under the same chemical gradients revealed the heterogeneity of chemotaxis responses in a genetically homogeneous cell population, especially the existence of a subpopulation with extremely high sensitivity for chemotaxis," and the title of their paper is "A sub-population of *Dictyostelium discoideum* cells shows extremely high sensitivity to cAMP for directional migration." Nonetheless, the authors do not cite the earlier work by Wang et al. (2012), and they do not specifically attempt to quantify the total population behavior in terms of the observed subpopulation thresholds. I can illustrate how to do that by applying the PBT approach to the data for the distribution of response thresholds across the entire population identified by Ohtsuka et al. (2021) (Fig. 4.5A). Empirical fitting easily identified two normal distributions whose sum closely matched the actual data (Fig. 4.5B). We can consider these as subpopulations 1 and 2, the former having a mean sensitivity of about 6 pM and a standard deviation of $10^{0.8}$ M, and the latter having a mean of 50 nM and a standard deviation of $10^{1.3}$. (Note that the distributions are on a log scale of cAMP concentrations, and so together they span an enormous range of concentrations.)

In addition, the microfluidic techniques employed by Ohtsuka et al. (2021) enabled them to measure the response times of individual cells to a given cAMP concentration. For a concentration of 10^{-7} M, the average response

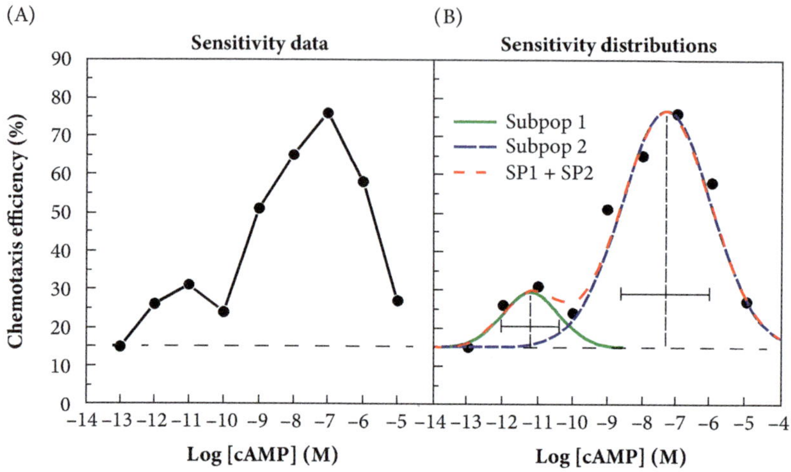

Figure 4.5 Sensitivity of *Dictyostelium discoideum* cells to movement in response to cAMP concentrations. **A.** The percentages of individual cells moving directionally through the parallel channels of a microfluidic apparatus in an all-or-none manner is plotted as the "chemotaxis efficiency" versus the cAMP concentration. **B.** The same data as in panel **A** (black circles) shown with normal distributions of two subpopulations. A more sensitive and smaller fraction of the total population (subpopulation 1; green curve) had an estimated mean sensitivity to cAMP of $10^{-11.2}$ M (6.3 pM) with a standard deviation of $10^{0.8}$ M. The larger and less sensitive fraction of the population (subpopulation 2; dashed blue curve) had an estimated mean sensitivity to cAMP of $10^{-7.3}$ M (50 nM) with a standard deviation of $10^{1.3}$ M. The sum of the threshold distributions of the two subpopulations is shown as the dashed red line, which closely matches the experimental data. One standard deviation on each side of the mean value is indicated for each distribution. **C.** The timing of cell movement in response to a cAMP concentration of 10^{-7} M. The predicted time courses for each cell subpopulation and their total are illustrated based on the population-based threshold model using the population distribution values from panel **B**. The time constants and the fractions of cells in each subpopulation were adjusted empirically to match the time course of the total population.

Experimental data points reproduced with permission from Ohtsuka, D., et al. (2021). A sub-population of *Dictyostelium discoideum* cells shows extremely high sensitivity to cAMP for directional migration. *Biochemical and Biophysical Research Communications* 554: 131–137. https://doi.org/10.1016/j.bbrc.2021.03.095.

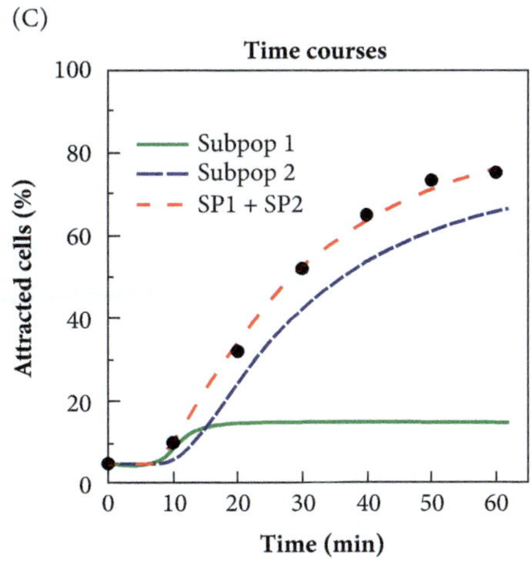

Figure 4.5 Continued

times to reach different percentages of the population are indicated by the data points in Figure 4.5C. Also shown in Figure 4.5C are the predicted time courses based on the PBT model parameter values of the two subpopulation distributions (Fig. 4.5B) by empirically choosing the time constant values and the fraction of the total responding population in each subpopulation. The thresholds of the highly sensitive fraction (subpopulation 1) would be exceeded by on the order of 10,000-fold at this test concentration, as indicated by their rapid response times and saturation of the response across the entire subpopulation (see the green curve in Fig. 4.5C). On the other hand, the cAMP concentration was only slightly above the mean threshold of subpopulation 2 (Fig. 4.5B). These cells responded more slowly, and not all the cells were attracted, as the test concentration did not exceed the thresholds of the entire population (Fig. 4.5B, C). When the time courses for the two subpopulations were added together (as they would be in the actual experiment, as the subpopulation that individual cells are in is not distinguishable), the time course matched the experimental data very closely (Fig. 4.5C). I will describe fitting subpopulations to the PBT model in more detail later (see Chapter 5 and the Appendix), but it is evident that the PBT model provides a simple, yet informative, approach to quantify multiple underlying sensitivity distributions (means and standard deviations) and that the kinetic aspect of the model (that the response speed increases in proportion to the extent that the factor level exceeds a given cell's threshold) also applies at the single-cell level.

Given the long history of study of chemotaxis in D. *discoideum* and other systems, it is rather surprising that population-based concepts are not commonly applied, particularly as they have been explicitly identified with variation in threshold sensitivities (Wang et al., 2012). Instead, models have tended to assume that the cellular machinery (e.g., the numbers of receptors present in a cell) are constant and have attributed different binding characteristics to different forms of the receptor molecules (Janssens and Van Haastert, 1987). This may, in fact, be the case, but the studies on receptor heterogeneity and cooperativity in cAMP binding were done on intact cells or isolated membranes, in both cases from large populations of cells, which may contain multiple distinct subpopulations. This is not to criticize the work that was done with the tools available, which discovered so much about this system, but it is instructive to appreciate the variation that can exist among individual protein molecules coded by the same gene (see Fig. 3.5) and add to that the variation among cells (Fig. 4.5A) that is uncontrolled in most experiments. For example, the numbers of binding proteins per cell have been estimated from apparent dissociation constants of different receptor protein forms, assuming that all of these forms are present in specific numbers in each cell (e.g., Mesibov et al., 1973). These estimates of mean receptor numbers per cell were used by Ohtsuka et al. (2021) to calculate that their highly sensitive cell population could detect as few as 6 cAMP *molecules* when the external concentration is 10^{-11} M, and other cells responded to 10-fold lower concentrations (Fig. 4.5A). It seems rather remarkable that a cell could sense less than one molecule of the signal (i.e., 0.6 molecules of cAMP). It would perhaps be easier to understand the high sensitivity of subfraction 1 if those cells simply increased the numbers of their receptor molecules by several orders of magnitude, increasing the likelihood that a cAMP molecule would encounter a receptor protein (Rodbard, 1973). On the other hand, the least sensitive fraction of the population might have relatively few sensor proteins, as a very high cAMP concentration (10^{-5} M) would still provide a positive signal as essentially every receptor protein would be triggered. Variation in sensor proteins would also readily explain how sensitivity adaptation could occur, through feedback mechanisms that increase or reduce the transcription, translation and sensitivity state (i.e., post-translational modification) of the receptor molecules. In a dynamic system actively responding to a given cAMP concentration, this would be virtually impossible to detect in pooled, extracted samples. However, we can readily imagine how this hypothesis could be tested experimentally by using reporters on the sensor proteins and microfluidic methods to identify individual cell responses and separate them into sensitivity subpopulations.

Ohtsuka et al. (2021, p. 136) also mentioned that "chemotaxis efficiency is sensitive to molecular noise" and that "[u]nderstanding the sensitivity of excitable systems to small signals and their robustness to molecular noise in individual cells will help clarify the heterogeneous sensitivity to small chemical gradients for chemotaxis." Biophysical models of chemotaxis have proposed "the possibility that 'noise' in the cellular signal perception/response mechanism can simultaneously account" for chemotactic properties of the signal reception components (Tranquillo et al., 1988, p 303). This may certainly be the case, and the biophysical properties operating at molecular scale must contribute to the macroscale properties observed (Parthasarathy, 2022). In general, however, these models envision homogeneity among the cells and receptors or at least homogeneity in their "noisiness." In contrast, we have seen that individual cells vary in their sensitivity thresholds across the entire response range. It would therefore seem reasonable that individual cells exhibiting a wide range of thresholds would be an essential component of the chemotactic or aggregation response, rather than simply an unavoidable consequence of stochastic noise, particularly across eight orders of magnitude of signal concentrations.

As noted previously, adjusting the overall receptor sensitivity as the cell moves up an attractant gradient is essential for the system to work. In fact, *D. discoideum* cells are somewhat larger (~5 nm in length) than bacteria and can actually sense a concentration difference of only 1–2% between the front and back ends of the cell to direct their movement. Yet this is independent of actual concentration across multiple log orders of magnitude (Mato et al., 1975; Parent and Devreotes, 1999). It is difficult to imagine a single "proteoform" of receptors that would exhibit such binding plasticity. It is much simpler to imagine that cells can shift their mean sensitivity with signal concentration in order to maintain precision in the relevant concentration range. As noted above, adjusting the number of receptor proteins per cell could automatically adjust the effective internal response level as the concentration changes. However, experiments in which large populations of cells are pooled together to measure receptor levels or properties will never be able either to detect or to test this hypothesis, as the proteins from all cells will be pooled together prior to bulk measurement. On the other hand, individual cells with fluorescent tags on the receptor proteins, when exposed to a range of signal concentrations, would be able to answer this question directly. It would not be efficient for cells to maintain a high sensitivity receptor complement that would be completely saturated (and therefore insensitive) at concentrations 100,000-fold higher. Appreciating this ability to vary sensitivity thresholds among individuals as needed, based on the current conditions, is much more stable and economical than imagining some stereotypical population of identical cells whose diverse behavior (which is critical to their survival) is fundamentally based on stochastic "noise" (Cortijo and Locke, 2020). That cells themselves can vary their sensitivities according to the conditions that they are currently experiencing is a much simpler hypothesis, and as we will see in multiple subsequent examples, this hypothesis is well supported by data (Cortijo et al., 2019; Cortijo et al., 2020). This ability to adjust and adapt is the essential advance of biology over the purely physical world in which individuals are identical and their behaviors are determined completely by random physical processes such as diffusion.

Quantal microbiology

In addition to receptor and signaling systems, microbial genetics, metabolism, physiology, growth and development all provide examples of interactions between variable individuals and (fairly) consistent population behavior (Brehm-Stecher and Johnson, 2004). This situation has been elegantly described as "quantal microbiology," which makes the analogy that "Classical physics is to quantum mechanics as classical microbiology is to quantal microbiology" (Bridson and Gould, 2000, p. 95). That is, Newtonian physics works well at the macroscale across large numbers of individuals; however, it breaks down when the behavior of fundamental particles is considered at their own spatial and time scales, where quantum physics must be applied (McFadden and Al-Khalili, 2014). Similarly, microbiologists generally study samples composed of 10^3 to 10^6 cells, tacitly assuming that these cells are essentially identical, and because large sample sizes are needed to provide sufficient material to analyze. This assumption is not unreasonable in cell cultures in which the environmental conditions are uniform and excess resources are provided. However, the existing variability and potential for variation in microbes is astounding. With respect to genetic variation among individuals, "a colony of 10^{12} cells may contain 10^{15} genes. With a background mutation frequency of 10^{-7}–10^{-8} or so, the single colony will contain 10^7–10^8 mutations at any one time" (Bridson and Gould, 2000, p. 95). Thus, even in so-called pure cultures, an enormous amount of genetic variation is present, and when subcultured, there is a significant element of randomness regarding which cells initiate growth and division first and eventually come to dominate the resulting culture. As noted previously, even if mutations are random, there are likely to be differences in the pathways through which they occur and in their consequences for growth of the cells. Thus, we can assume that this underlying genetic variation is present in population distributions, their relative abundances representing the likelihood that a specific combination of mutations would be present in a given cell. In uniform environments in which competition for resources is minimized, we can expect that the likelihood of a given cell succeeding would be related to its prevalence in the population, which is higher near the mean of the population distribution. This is fortunate for microbiologists, as it enables them to maintain specific cell lines in culture without exhibiting the extreme variation that is present.

In their native environments, however, resources are not uniformly distributed, and the inoculum initiating a particular colony's growth can be much more random. In other words, "Large numbers of organisms can give predictable results but small numbers give unpredictable results" (Bridson and Gould, 2000, p. 97). (See a similar comment with reference to seeds presented in the Conclusion of Chapter 5.) This effect of environmental variation was demonstrated experimentally by engineering a bacterium that grows on plant leaves (*Erwinia herbicola*) to express a stable detectable marker, green fluorescent protein (GFP) (Remus-Emsermann and Leveau, 2010). Production of this protein could be induced in a culture of bacterial cells, and then the inducer could be removed, leaving a fixed amount of the marker protein present in each cell (which also varied in a normal distribution in the cell population). As the cells subsequently divided, the initial amount of the GFP protein would be split between the daughter cells, reducing the signal per cell by half. This amount would reduce by half again in the next cycle of division, and so on. This process could be continued for four to six generations until the signal was too diluted to detect. As expected in cultured cells, this dilution progressed at a similar rate among all cells in the population; that is, essentially all cells were able to grow and divide at a similar rate. However, when the same cellular populations were inoculated onto a plant leaf, there were wide variations in the rates of loss of the marker signal, indicating correspondingly large differences in the cells' access to favorable environments for growth and division or in their ability to take advantage of the resources present. It appears, therefore, that not only is there large inherent variation among the bacterial cells, but also "the fate of individual immigrants is determined in large part by the environmental heterogeneity at the microscopic leaf level" (Remus-Emsermann and Leveau, 2010, p. 219). Thus, the biological reality is the continuous interaction of variable populations of individual organisms with heterogeneous or "patchy" environmental circumstances.

As Leveau et al. (2018, p. 2) noted, "One major discovery in recent years is the existence and extent of heterogeneity among [individual microbes]." This field has grown along with the development of tools and technologies for studying microorganisms individually at their own scale and in their own environments (Brehm-Stecher and Johnson, 2004; Richards et al., 2019; Muljadi et al., 2022). Grujcic et al. (2022, p. 1), noted, "One emerging trend is the unfailing observation of heterogeneity in activity levels within microbial populations. Wherever researchers have looked, intra-population variability in biochemical composition, growth rates, and responses to varying environmental conditions has been evident." An understanding of this intra-population variation, however, is still evolving. As discussed above, initially it was considered to be "noise" associated with stochastic processes around an "average" cell. However, as experimental and statistical analysis procedures have progressed, it is becoming appreciated that "a purely stochastic representation of cellular heterogeneity is not appropriate for a large proportion of genes and biological processes" (Llamosi et al., 2016, p. 2). Instead, these authors explored "a different approach in which variability is represented only as stable differences among cells." In addition (p. 12), "the information on each and every cell is jointly used to calibrate the model parameter distribution, alleviating the problem of limited observability and noisy observations encountered at the individual-cell level." Using this approach, Llamosi et al. were able to model the transcriptional behavior of large numbers of yeast cells in response to hyperosmotic shocks in terms of distributions of cells exhibiting specific behaviors, and they showed that the "average" behavior did not coincide with the overall population means.

There is now a growing recognition that biological variation is present in populations of individual microbes and that this represents true variation, not simply experimental error. Biochemical pathways highly homologous to those present in microbes are also present in many critical systems in multicellular organisms (Koshland, 1980), including those involved in development, hormonal responses and adaptation to environmental stresses (Falke et al., 1997). Methods have been developed to conduct experiments on a cell-by-cell basis in multicellular organisms. For example, molecular reporters associated with specific cellular processes can be observed microscopically in many cells simultaneously and then synchronized (such as for cell division times) *in silico* to provide large datasets revealing cyclic cell behavior within unsynchronized populations (Cappell et al., 2018). Statistical methods are also available to identify populations and derive their means, variances and other characteristics (Llamosi et al., 2016). These methods overcome some of the inherent difficulties in working with large numbers of individual cells, and they also apply to plant and animal cell cultures. Even with these sophisticated techniques, the variation observed generally is attributed to genetic variation, to lack of synchronization, or to stochastic chemical processes, rather than to the possibility that the inherent variation among individuals may constitute a biological regulatory system. The two features of the PBT model—recruitment of additional individuals and more rapid kinetics as signaling factors increase—provide a mechanistic rationale for the existence and maintenance of such variation. Often lacking in these studies are attempts to determine specifically the sensitivity characteristics of these populations to different factors (e.g., Fig. 4.5), particularly signaling molecules or hormones, which often require changes in methodology. As Altschuler and Wu (2010, p. 561–562), noted:

> Determining whether observed heterogeneity has functional significance requires a framework for quantifying heterogeneity and assessing its information content. An intuitive and tractable approach is to decompose heterogeneous populations into mixtures of simpler, more homogeneous subpopulations, based on the expectation that cells with relatively similar measured states should behave in a similar way. The biological relevance of such a decomposition could be tested by determining whether different subpopulations, or mixtures of subpopulations, have different functional properties. . . . In general, the investigation of heterogeneity requires a combined approach for capturing population statistics from single-cell measurements, identifying patterns of heterogeneity, and testing whether these patterns contain functional information. . . . The ability to transform heterogeneity into a tractable, computable, and bone fide property of cellular populations provides a rigorous starting point for determining which variation is random and which is meaningful, and for building better models of the behaviors of individual cells.

This information, gained largely from dose-response experiments to measure the sensitivities of individuals to doses of specific factors, is needed to fully test the applicability of the PBT model to these systems.

Microbiomes, pathogens and immunity

Thus far in this chapter, microbial variation has been primarily addressed in relation to specific species or genotypes. However, microbes in the natural world do not exist in isolation, but rather in microbiomes containing multiple diverse species, as is evident in the vast microbial diversity present in oceans, soils, plants and animals (McDonald et al., 2020). It is beyond the scope of this book to delve deeply into this important and expanding field, but given the population-based approach of the PBT model, there is little doubt that it could be applicable to the study of interactions within and among microbiotas. A clear example of this is the work of Metcalf and Koskella (2019) in describing the interactions of commensal (nonpathogenic) and pathogenic microbiota in relation to immune responses by host organisms. They illustrate population distributions of these microbiotas and what happens when they overlap in relation to a variable threshold for triggering an immunity response (Fig. 4.6). Depending upon the extent of overlap of the populations, their relative sizes, and the sensitivity thresholds of immune response mechanisms, different subpopulations can be identified. As the authors explain (p. 536):

> The problem faced by the immune system in discriminating between "good" commensals and "bad" pathogens is analogous to a classic challenge from epidemiology. Overlap between two categories along a continuum means that choosing where to draw a line to discriminate between them results in a trade-off between falsely allocating individuals to one or the other category. In this example [Fig. 4.6], the two distributions correspond to the community of commensal microbiota (left)

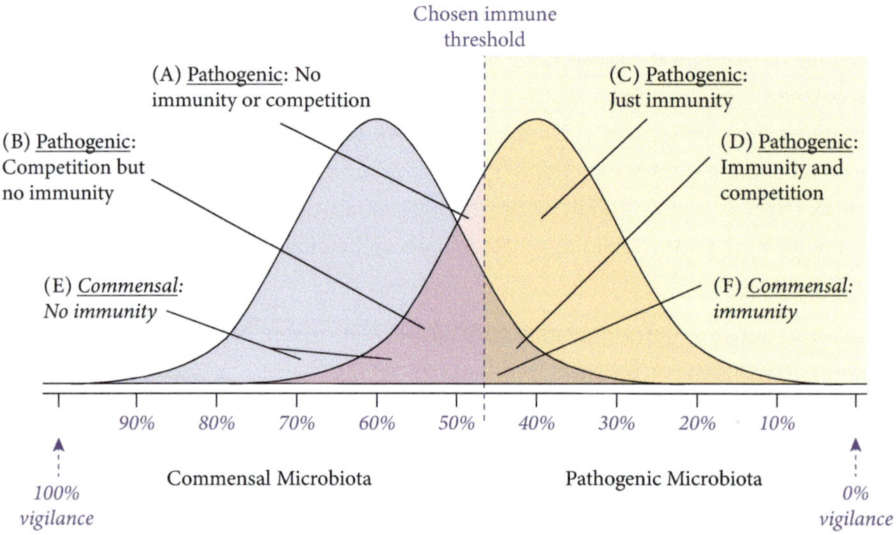

Figure 4.6 Diagram of overlap between nonpathogenic commensal microbiota (left, blue) and pathogenic microbiota (right, brown) along a continuous axis of increasing intensity of an immune system trigger. The purple dotted line shows one possible location for a threshold for triggering an immune response. As illustrated, the extent of population overlaps and locations of their thresholds result in different sectors that can be identified as labeled. Not shown is the likely variation in the abundance of the two populations, affecting the heights and widths of their distributions. The biological optimization of immunity responses involves where to place the threshold to balance the costs of the pathogenic microbiota and immunity responses versus the benefits of the commensal microbiota.

Reproduced with permission from Metcalf, C.J.E., and Koskella, B. (2019). Protective microbiomes can limit the evolution of host pathogen defense. *Evolution Letters* 3(5): 534–543. https://doi.org/10.1002/evl3.140.

> and pathogenic microbiota (right); the height of the two distributions reflects the number of individuals in each category corresponding to each value along the continuum (e.g., shared flagellin characteristics). The immune strategy under selection corresponds to the problem of where to draw the line above which an immune reaction is triggered. If we define the commensal microbiota as the community of microbial species with positive effects, ideally, the line would have all individuals from this community to its left, where no immune response is triggered, and all pathogenic microbiota to its right, where immunity acts to clear microbial species. . . . The strategy that maximizes fitness will balance costs of immunity with costs and benefits of pathogens and commensal microbiota, and the effect of their overlap, that is, the impact of the commensal microbiota on the pathogen relative to outcomes for the host.

If, as argued in this book, much of biology actually consists of overlapping populations of individuals with related but variable activities across levels of complexity, we can envision that the approach described here could be applied to understand how overall function could be optimized on both physiological and evolutionary timescales by adjusting multiple population distributions in relation to their triggering thresholds. This will only be enhanced by recent developments allowing single-cell transcriptome analysis of microbial communities (Pountain and Yanai, 2025).

Conclusions

Recent research on microorganisms has provided a wealth of evidence for the existence of individual variation at the cellular level, based on both genetic and physiological variance. Available dose-response data often closely follow the predictions of the PBT model, including the opportunity for combinations of distinct subpopulations to represent the overall population behavior. Microbial scientists are ahead of organismal scientists in this regard, with the former already having defined "microbial individual-based ecology" or "quantal microbiology" as a key subdiscipline. The increasing recognition of the critical importance of individuality to microbial behavior is matched by the extension of this analysis to the microbiome level. The key aspects of the PBT approach are all present: inherent variation among individuals in their sensitivities and responses organized into population distributions, recruitment to respond in proportion to signaling factor concentrations, and kinetic (response time) behavior consistent with signal levels above individual sensitivity thresholds. Thus, microbiology as a field seems ripe to exploit the simplicity and integrative potential of the PBT approach.

5
Plants

Plant individuality and behavior

If we think broadly about plants, they will include those organisms capable of using photosynthesis to capture energy and store it in chemical compounds. The earliest organisms with this capability were cyanobacteria, single-celled prokaryotic organisms that are not generally considered to be plants. Nonetheless, by about two billion years ago, their generation of oxygen as a product of photosynthetic activity had initiated oxygenation of the earth's atmosphere (Luo et al., 2016), eventually making it capable of supporting the respiratory needs of more advanced organisms. Endosymbiosis of a cyanobacterium by a eukaryotic protist resulted in chloroplasts (the organelles in plant cells responsible for photosynthesis) and development of the plant evolutionary lineages. Single-celled plants such as *Chlamydomonas* have been widely studied as examples of early eukaryotic photosynthetic organisms (Merchant et al., 2007). As noted earlier in discussing bacteria and fungi, these single-celled organisms had evolved sophisticated sensing and response capabilities that are found in subsequently evolving photosynthetic organisms that emerged from the seas and colonized the land surface. This evolution required new cellular and tissue-based mechanisms to manage water uptake (i.e., roots and water transportation vessels) and loss (hydrophobic surfaces and leaf pores called stomata to limit and control water loss) in a terrestrial environment. These capabilities range from the bare minimum, as in mosses, to trees capable of growing to 100 m heights and living for several thousand years.

The colonization of land by plants enabled the subsequent emergence of animals from the sea by providing them a terrestrial food source. This eventually led to the evolutionary explosion that resulted from the interactions of flowering plants and pollinating insects that we enjoy in the diversity of flowers in our gardens, parks and wildlands. In addition, as early land-dwelling plants spread, they encountered much greater variation in environmental conditions than occurred underwater. In many ways, the evolution of plants is a record of the varied ways that the photosynthetic apparatus can be

Populations of Individuals. Kent J. Bradford, Oxford University Press. © Kent J. Bradford (2026).
DOI: 10.1093/9780197846032.003.0005

organized and packaged (in both space and time) to support an organism's growth and reproduction in a specific environment. This requires access to light, water and CO_2 gas and is subject to the economic constraint that accessing CO_2 from the air costs the plant water through evaporation. In addition, the earth's tilted orbit around the sun imposes seasonal temperature constraints on plants (and animals), not to mention the diurnal presence and absence of light, the essential driver of photosynthesis, due to the earth's daily rotation.

It may therefore be concluded that a critical component of the distribution of plant life on land involved their ability to sense their environment and to respond to it in an adaptive manner to maintain photosynthetic productivity. This adaptive capacity underlies the application of terms such as "intelligence," "behavior" and "agency" to plants (Trewavas, 2014; Gilroy and Trewavas, 2023), which lack the nervous system that we generally associate with sentience, consciousness and agency in animals. We continue to discover the diversity of mechanisms that plants use to sense and adapt to their environments across all space and timescales, from the acquisition of nutrients by roots at the molecular scale to the morphological and physiological adaptations that enable plants to grow in all but the most extreme environments (i.e., the polar regions and the highest mountaintops). Fritz et al. (2018) provide an in-depth overview of this issue with respect to plant leaves and their plasticity in response to environmental variables. It is in this context that I will try to describe the roles of sensitivity thresholds and their population distributions in some of the key features of plant life.

Single-celled algae

Chlamydomonas reinhardtii is a unicellular green algae that has been present on earth for over a billion years (Blaby-Haas and Merchant, 2019). It has been widely used to study eukaryotic photosynthesis, as it can be grown in culture, enabling experimental control and manipulation of its growth conditions (Merchant et al., 2007). As with many studies of microbes (such as yeast), growth conditions are often maintained relatively constant (i.e., constant light) and the cells individually progress through their growth and division cycles. This process has generally been thought to result in unsynchronized populations in which different fractions of the total cell population will be at different stages of this cycle, as in yeast. Such "log phase" cultures are rapidly increasing in cell numbers (i.e., exponentially, as each cell that divides creates two cells); thus, in a sense any sample is assumed to be randomized with respect to developmental stages. However, it was discovered recently that even when grown in constant light and nutrient conditions, *C. reinhardtii* cells exhibited an ongoing diurnal (or circadian) cycle in their transcriptional and metabolic activities (Salomé and Merchant, 2021). This cycling can be entrained and enhanced by exposing the cells to light on a diurnal basis (12 h of light and 12 h of dark) (Strenkert et al., 2019). The cells undergo massive changes in transcription and protein synthesis to maximize photosynthesis and growth during the light period and then use the accumulated photosynthetic resources for division followed by relative quiescence during the night. This process is also apparent in higher plants, which are also entrained to circadian rhythms of light or temperature (Salome et al., 2008).

While this feature of plant life has been recognized for many years, as in the daylength requirements for flowering among some plant species adapted to higher latitudes (Gendron and Staiger, 2023), it has a consequence that has only recently been explicitly considered in experimental designs. As shown for *Chlamydomonas* (Strenkert et al., 2019; Ma et al., 2021) and for the experimental model plant *Arabidopsis thaliana* (Hsu and Harmer, 2012), failure to recognize diurnal cycling can lead to misleading results when analyzing transcriptomes. Even as little as 30 minutes difference in sampling times can result in significant differences in the abundance of transcripts associated with specific diurnal stages (Hsu and Harmer, 2012). In fact, once this fact is recognized, the appearance and disappearance of specific co-expressed genes can be used to artificially synchronize events occurring in unsynchronized cultures (Ma et al., 2021, p. 1043): "Indeed, the clustering of cells that fall along a continuum such as a developmental time course more accurately represents the underlying structure than discrete samples." Similar approaches can also distinguish among different cell types during nematode development, which are obscured by bulk tissue sampling methods (Ma and Zheng, 2023). The consequence of bulk sampling methods is stated clearly by Ma et al. (2021, p. 1042): "bulk RNA-seq data represent an average of typically hidden heterogeneity in the population." The importance of population-based thinking and experimental approaches in order to uncover this hetergeneity is evident.

Plants almost universally exhibit diurnal cycling of their transcriptomic, proteomic and metabolic processes. Thus, "it is important to take the time of day into account when designing all types of large-scale experiments" (Hsu and Harmer, 2012, p. 10). Primary emphasis in transcriptomic studies is often placed on the differentially expressed genes (DEGs). Exceptional statistical energy is often directed toward demonstrating the significance of even relatively minor differences in expression of specific transcripts in pooled, bulk samples (generally tissue samples containing multiple cell types). However, in circadian organisms (which also include animals), simply changing the time of day of sampling can result in much larger changes in expression of major cohorts of genes. In fact, these cohorts, often associated with developmental or diurnal changes in metabolic needs, represent subfractions of the total transcriptomic capacity that are expanding and shrinking in a cyclical manner. A population-based approach would recognize the sensitivity of these transcriptional subfractions to their environmental conditions and analyze their waxing and waning as changes in signal levels or in the sensitivities of their promoters to the relevant signals (Fig. 5.1) (Strenkert et al., 2019). Consistent with the PBT approach, also included would be both recruitment (more genes in a family) and rate (more rapid transcription of genes in the cohort) as the signals and sensitivity to them cycle up and down through feedback mechanisms driven by the cellular circadian clock. Once these patterns are identified and quantified through single-cell approaches, it is possible to interpret the results of larger scale or bulked samples by understanding the underlying population-based parameters and the kinetics of their daily cycles.

Chloroplast movements

As noted above, chloroplasts are the organelles in plant cells that conduct photosynthesis, or the capture of CO_2 from the air and its conversion into carbohydrates, from which all other components in plants are derived. Chloroplasts were originally single-celled cyanobacteria that were incorporated as organelles into eukaryotic cells, retaining some of their original bacterial genome, but now also requiring some components encoded in the plant's nuclear genome (Ku et al., 2015). As organelles, chloroplasts have retained the bacterial ability for movement but use a different mechanism from the flagellar one described in Chapter 4. For example, it has been known for more than a century that chloroplasts adapted to dim light will move within the cell to avoid high light intensities (Wada and Kong, 2018; Fig. 5.2). This movement is accomplished by specific receptors (CHUP1 in Arabidopsis) that interact with the actin filaments in the cell cytoskeleton and allow the chloroplasts to jump. Otherwise, their movement is localized and random due to diffusion or undirected attachment to the actin filaments. They can direct this adaptive movement by shifting the binding proteins to different locations on the organelle, which increases the velocity of movement in the direction of the organelle that contains more of the binding proteins (Gotoh et al., 2018; Wada and Kong, 2018; Kong et al., 2024; Suetsugu and Wada, 2020).

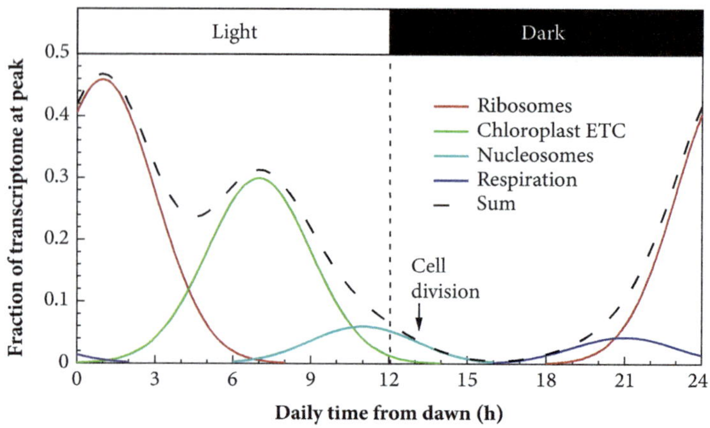

Figure 5.1 Illustration of the diurnal cycle of key transcriptional cohorts in *Chlamydomonas reinhardtii* cells entrained to a cycle of 12 h light and 12 h dark. Ribosomal protein expression increases prior to dawn in anticipation of synthesizing chlorophyll and nuclear-encoded proteins of the chloroplasts (chloroplast ETC [electron transport chain]). Nucleosomes increase prior to cell division, which occurs within the first hour after darkness. Respiration-related genes increase during the night as energy production from light is absent. Many such cohorts that cycle diurnally are identified, but individually they represent too small a fraction of the total transcriptome to be shown here. It is evident, however, that the composition of the transcriptome varies radically during the course of a day, making the timing of sampling critical for defining differentially expressed genes. In addition, representing these cohorts as populations that increase and decline daily provides a clear way to visualize their patterns by varying their magnitude (number of genes/transcripts) and standard deviation (time span over which they increase or decrease).

Adapted with permission from Strenkert, D., et al. (2019). Multiomics resolution of molecular events during a day in the life of *Chlamydomonas*. *Proceedings of the National Academy of Sciences USA* 116(6): 2374–2383. https://doi.org/10.1073/pnas.1815238116.

Is this movement of chloroplasts analogous to chemotaxis? In fact, some high-sensitivity individual chloroplasts detect a change in light intensity and react quickly. There is also some coordination among the chloroplasts, so those early responders could trigger additional chloroplasts to move, as in quorum sensing. The description by Kagawa and Wada (2004, p. 594) of chloroplast perception and response compares closely to that of chemotaxis:

> Signal concentration differences could be sensed by the chloroplasts as both temporal and/or spatial changes. If chloroplasts move, they can compare the signal concentration between two different locations, temporally and spatially. However, when high fluence rate [light intensity] is applied, chloroplasts show avoidance response without vacillation. Thus, chloroplasts may be able to detect a temporal

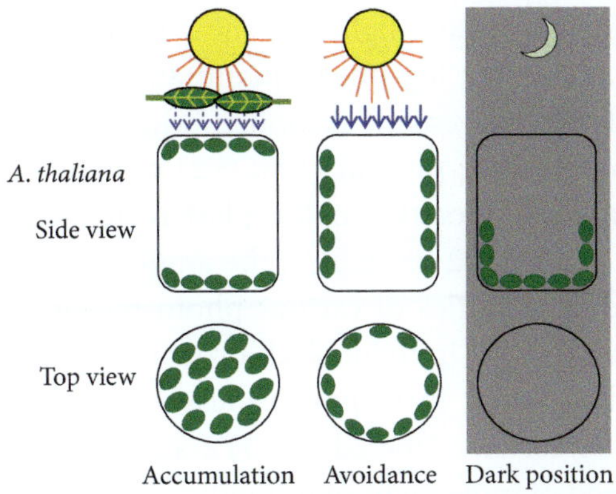

Figure 5.2 Chloroplast movements in *Arabidopsis thaliana* cells. In weak light conditions under the canopy or in shade, the accumulation response occurs, and the chloroplasts spread horizontally across the upper and lower cell walls to maximize light interception (left). In strong light conditions, as under direct sunlight, the chloroplasts exhibit an avoidance response in which they migrate to the vertical cell walls, reducing their exposure to intense light (center). In darkness, the chloroplasts tend to position on the lower side of the cells (right).

Reproduced with permission from Wada, M. and Kong, S.G. (2018). Actin-mediated movement of chloroplasts. *Journal of Cell Science* 131(2): jcs210310. https://doi.org/10.1242/jcs.210310.

> difference in signal concentration over the time course of a microbeam exposure. Another explanation may be that a single chloroplast covers a significant area to detect a difference in concentration of some signal between two spots of the chloroplast closer to and farther from the beam border. When the difference in levels of this signal concentration between the two areas in a single chloroplast is high enough, information regarding the rate of avoidance, as well as directional orientation, could be perceived.

If the analogy with chemotaxis holds, we could expect population variation among the chloroplasts in sensitivity to light intensity. In fact, Schramma et al. (2023) proposed a threshold mechanism and model for this movement. They followed individual chloroplasts moving in response to a change in light intensity and showed that some fractions of the population are more sensitive and will jump away more quickly. Eventually all of them move to the sides of the cell away from the direct light. In their model, they also show wide normal distributions of responses in the population. In fact, they use a "white noise" Gaussian model underneath it, basing the actual mechanism primarily on diffusion. However, Schramma et al. do not consider the possibility that different chloroplasts have different sensitivities and therefore respond at different rates to a given change in light intensity. Only dim light and high intensity white light are shown for a dose-response curve. In fact, studies have shown log-linear quantitative responses to light intensity in this avoidance movement response, which is dependent upon the amount of a photoreceptor protein (PHOT2) present (Kagawa and Wada, 2004, p. 594): "Since the velocity of chloroplast avoidance movement appears to be dependent on PHOT2 levels . . . , the amount of PHOT2 may change the threshold and sensitivities of various photo-responses." Thus, variation in the PHOT2 content of individual chloroplasts could contribute to the population-based variation in responsiveness; this could be tested experimentally using fluorescence-labeled PHOT2 and monitoring its content in relation to the speed of movement of individual chloroplasts.

The binding protein-actin movement model indicates that it is under control of the chloroplast, and it would be simple for the numbers of receptors such as PHOT2 to vary among the organelles. This could account for adaptation (adjustment of numbers to light intensity) and speed (the early jumpers could have more receptors) of the avoidance response. Schramma et al. (2023) include response thresholds in their model, but they do not consider that the chloroplasts themselves could be the basis of variation, despite showing Gaussian distributions of velocities among individual organelles. Nor do they consider that individual chloroplasts differentially jump under a given condition. The authors argue instead that this behavior is due to a physical change between a glassy versus fluid state in the cellular medium influencing diffusion rates. A physical change like this is a possible contributor, but a normal distribution of photosensitivities among the chloroplasts could be the underlying source of the variation, rather than just being considered to be part of a background random "noise" function required to fit their model. Schramma et al.'s data also showed logarithmic variation among individuals exhibiting differing thresholds and speeds of response, as is often identified by the PBT model.

Plant development

Development occurs continuously and repeatedly during the plant life cycle and is also sensitive to both current and past environmental conditions. For example, leaves that develop in low light ("shade" leaves) are morphologically, anatomically and cellularly different from leaves developing in high light ("sun" leaves) even on an individual plant (Fritz et al., 2018). For deciduous plants that lose their leaves each fall, this process is repeated every spring, with new stems and leaves growing from dormant buds and adapting their phenotypic characteristics to their current environment during their development. Plant roots detect and grow toward sites in the soil with more plentiful water and nutrients. The growth of water-carrying vessels (xylem) in the stem depends upon the conditions (particularly water availability) during their production and is the basis of variable tree rings that can be used to track past climatic conditions (dendrochronology). When induced to flower, cellular development is switched from making leaves to making floral parts. This unique ability of plants to continuously develop (and adapt to current or anticipated conditions) is due to the flexibility of their meristems, which are the sites where cells divide and subsequently progress through different developmental pathways to form all parts of the plant (Kean-Galeno et al., 2024). However, while the developmental processes from cells to organs exhibit robustness in final organ outcomes, fundamental heterogeneity is present in the underlying cellular processes such as cell division, expansion and differentiation (Hong et al., 2018; Marand et al., 2025). As Hong et al. (2018, pp. 479, 480) noted, "The plant appears to be exploiting stochasticity in expression to generate heterogeneity in cell size," and therefore "the plant must have mechanisms to ensure robust organogenesis from heterogeneous cells."

Hormonal regulation

While much remains to be learned, it is clear that plant hormones, particularly auxin (indole-3-acetic acid), are involved in organizing these meristems and subsequent tissue and organ development (Martin-Arevalillo and Vernoux, 2023; Liu et al., 2024). The particular arrangement of leaves around a plant stem, termed phyllotaxis, is initially set based upon the relative positions of their initiation in the meristem (Tian et al., 2020), which is controlled primarily by auxin (Kean-Galeno et al., 2024). Subsequent cell expansion to the leaf's final size is also dependent upon auxin (and other plant hormones). Auxin also is mobile, moving both by diffusion and directionally via asymmetrically arranged transporters in the cells' plasma membranes, called PIN proteins, and via plasmodesmatal cytoplasmic connections between adjacent cells (Martin-Arevalillo and Vernoux, 2023). Such directed movement enables polarity in cell growth and development and led to the concept that auxin concentration gradients were a primary component of its action. Of course, for a hormone to act, it must be present at a concentration sufficient to trigger a response, but its effects also are dependent upon specific target cells being receptive to it, indicating that sensitivity to the hormone must also be involved. Four decades ago, there were brisk arguments about whether auxin responses were controlled by the amounts of auxin or the sensitivity of various cells to the hormone (Trewavas and Cleland, 1983). This system was an early subject for interpretation in light of the PBT model, which resolves the conflict by envisioning a distribution of sensitivity thresholds among individual cells, which then respond when the hormone concentration at their location exceeds their response thresholds (Bradford and Trewavas, 1994; Trewavas, 2012; Bradford, 2018; Watahiki and Trewavas, 2019).

The important role of hormonal sensitivity in directing development has been detailed in studies of how the top (adaxial or dorsal) and bottom (abaxial or ventral) sides of leaves develop during initiation of a new leaf at the shoot meristem (Nakayama et al., 2022). This dorsiventrality of leaf development is characterized by a sharp distinction between the adaxial and abaxial sides of the developing leaf from very early in its initiation in the meristem. The adaxial side generally has a thicker epidermis cell layer with a waxier surface, while the abaxial side often has more stomatal pores for entry of CO_2 (see below). However, somewhat surprisingly, this cellular differentiation initially occurs at a relatively high, but constant, auxin concentration and is driven by prepatterning of the cells by specific transcription factors that influence auxin sensitivity by regulating the expression of auxin response factor (ARF) proteins (Burian et al., 2022). In addition, regulatory small RNA molecules (see Chapter 3) also participate in this process (Skopelitis et al., 2017, pp. 265, 266):

> Small RNAs possess intrinsic properties that make them particularly well suited to drive developmental change. These include a high degree of specificity (particularly in plants), a direct and rapid mode of action, and the ability to confer sensitivity and robustness onto gene regulatory networks (Schmiedel et al., 2015; Plavskin et al., 2016). Moreover, small RNAs display marked versatility in the regulatory effects they have on their targets. For instance, miRNAs can clear target transcripts to delineate domains of accumulation or act as rheostats by dampening noise in target gene expression (Skopelitis et al., 2012). . . . [T]he concentration gradients formed by mobile small RNAs create sharply defined domains of target gene expression that underlie robust developmental boundaries. This is mediated by a non-linear, threshold-based readout that is highly sensitive to relative[ly] small RNA-to-target levels, allowing target gene expression boundaries to be flexibly positioned. Intriguingly, these patterning properties are fully intrinsic to the small RNA gradient itself and do not rely on feedback regulation from downstream network components. Thus, while sharing conceptual similarities with classical morphogens described in animal systems, mobile small RNAs constitute a

distinctive direct mechanism through which to relay positional information and drive developmental change.

These small microRNAs (miRNAs) influencing auxin responses also move within the developing tissue, creating their own concentration gradients (Skopelitis et al., 2017). When the miRNA concentration is high relative to its target mRNAs, the response becomes an off switch, while when the concentration is low relative to the target population, miRNA concentration becomes more like a rheostat, scaling expression relative to the miRNA/target ratio (Skopelitis et al., 2017; Fig. 5.3). This threshold-like behavior contributes to a sharp boundary between adaxial and abaxial cell development in the leaf.

The combinations of contributors, including auxin, transcription factors, response factors and small RNAs, interact in positive and negative feedback loops to result in the consistent bilateral developmental pattern as leaf formation progresses. This has been described computationally and tested experimentally on the basis of a self-organizing Turing reaction-diffusion system by Scacchi et al. (2024, pp. 413, 419):

Here we harness the power of theoretical and computational models, molecular genetics and live confocal imaging to demonstrate that an organ-autonomous small-RNA-based Turing reaction-diffusion system coordinates the spatial

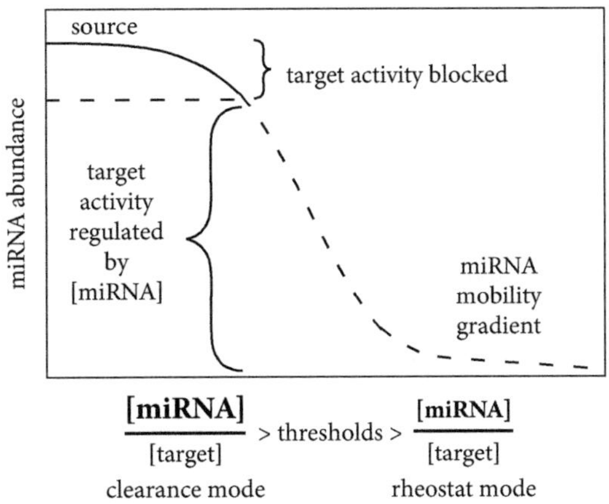

Figure 5.3 Threshold-type on/off behavior at high miRNA abundance transitions into analog (rheostat) regulatory mode at lower miRNA abundance in the determination of leaf dorsiventrality development.

Adapted with permission from Skopelitis, D.S., et al. (2017). Boundary formation through a direct threshold-based readout of mobile small RNA gradients. *Developmental Cell* 43(3), 265–273. https://doi.org./10.1016/j.devcel.2017.10.003.

> organization of adaxial and abaxial domains in developing leaves. Our integrated approach shows that the Turing dynamics when transiently guided by prepatterned information from the meristem environment is sufficient to self-sustain a correctly oriented bipolar pattern during growth. The system displays a remarkable adaptability and robustness, explaining the propagation of a stable planar polarity boundary in flat leaves, while simultaneously affording flexibility to accommodate developmental and evolutionary diversity in organ form. Thus, our findings identify a unique small-RNA-based Turing network as a dynamic regulator of organ polarity that accounts for leaf shape diversity at the level of the individual organ, plant or species. . . . Through cell-cell communication and feedback regulation, the polarity GRN [gene regulatory network] functioning in each cell orchestrates a collective cell behaviour that continuously organizes the spatial arrangement of ad–ab [adaxial-abaxial] domains at the organ level. As such, ad–ab gene expression can dynamically adapt to internal and external perturbations to sustain a robust polarity boundary in planar leaves while providing the flexibility needed to support morphological diversity.

The Turing system referred to here describes outcomes when two diffusible substances interact with each other to generate a spatially periodic pattern (Turing, 1952). This Turing analysis has been applied to developmental morphology to understand the formation of stripes on zebras and spotting patterns on giraffes, for example. It is now being applied successfully to the self-organizing patterns of plant meristems, as shown above. However, Scacchi et al. (2024, p. 413) note that it works "when transiently guided by prepatterned information from the meristem environment." From a PBT point of view, this "prepatterned information" could be interpreted as the distribution of sensitivity thresholds for the receptor components of these interacting systems. As with the original controversies about the basis of hormonal action, both the concentration and the sensitivity to it at a specific location and time are factors in the biological outcome. The Turing system approach nicely models the complex interactions of the mobile components of the signal-response system, but these mobile components interact with other system components generated in the responsive cells, such as the target mRNA abundance, as illustrated in Figure 5.3. I would suggest that adding a cell population sensitivity distribution component to the Turing models of auxin/miRNA distributions might be a useful addition to the solely concentration-based models. A specific gene (*EPFL2*) has recently been identified that controls cellular sensitivity to auxin in shoot cells (Tameshige et al., 2025), enabling more specific experiments to investigate the roles of production, movement, concentration and sensitivity of EPFL2 proteins in determining overall auxin effects. The components controlling sensitivity might also have their own network relationships, resulting in distinct models describing the concentration components and the sensitivity distributions determining the overall response (see Fig. 7.2). Combining these could be complex, but our models must be as complex as biological reality if they are to explain it.

Another intriguing aspect of auxin regulation of plant development is its involvement in many diverse tissues and growth processes. In addition to contributing to meristem organization and leaf development, auxin is also involved in cell division and growth, formation of lateral roots, and many other aspects of plant development and function (Cohen and Strader, 2024). However, this broad functionality of auxin raises additional questions of how a single, relatively simple chemical can play such a diversity of roles in different plant tissues. One likely possibility is that different auxin response factor (ARF) proteins are involved in different functional responses to auxin. The Arabidopsis genome contains 23 ARF transcription factor family members, each responding to auxin but potentially controlling a distinct downstream gene expression network under different spatial, developmental or hormonal conditions. In addition, after translation, ARF proteins can also be active or inactive through the formation of liquid–liquid condensates in the cytoplasm (Powers et al., 2019; Jing et al., 2022). Specific amino acid regions in proteins can enable them to enter these cellular membrane-less compartments, sequestering them and preventing their action, but also enabling them to reenter the active cytoplasmic phase when conditions change (Emenecker et al., 2021; Field et al., 2023). This provides a mechanism for adjusting cellular sensitivity to auxin in real time. As conditions change, such as diurnally or in response to specific developmental or environmental conditions, ARF proteins can readily shift between active and inactive liquid–liquid phases. This provides a mechanism for the sensitivity of a cell to auxin concentration to be adjusted in response to current conditions within an individual cell without requiring protein degradation or resynthesis after transcription. This further reinforces the concept that the sensitivity of cells to auxin (or other hormones) is not fixed but rather varies among subpopulations of cells and even in an individual cell over developmental or diurnal time. It also is consistent with the recruitment aspect of the PBT concept, as it provides a mechanism for a reservoir of ARF proteins to be present, but a varying fraction of them could be maintained inactive in the condensates (Klosin et al., 2020; see Fig. 8.2). Changes in auxin concentration could potentially add to or withdraw from this reservoir of ARF molecules, providing the kinetic and recruitment components of the PBT model that are particularly evident as hormone concentrations exceed the minimum thresholds for a response.

Without going into great detail, it is worth noting that photobodies (PBs) are nuclear condensates that contain the photoreceptor and thermosensor protein called phytochrome B (PHYB) and are associated with light and temperature perception in plants (Chen and Chory, 2011). Using methods to distinguish differently tagged PHYB proteins, Du et al. (2024) identified at least a dozen distinct populations of PBs that appeared or disappeared in response to changes in temperature. These distinct PBs types were not random, but rather were associated with specific locations in the nucleus and with their different thermosensitivities. Du et al. (2024, pp. 1, 9), found that:

> Intriguingly, warm temperatures reduce PB number by inducing the disappearance of specific thermo-sensitive PBs, demonstrating that individual PBs possess different thermosensitivities. These results reveal a nonrandom PB nucleation model, which provides the framework for the biogenesis of spatially distinct individual PBs with diverse environmental sensitivities within a single plant nucleus. . . . Our results suggest that the control of PHYB condensation nucleation could represent a previously uncharacterized mechanism in environmental sensing. This study paves the way for investigations of the mechanism of PB nucleation and the functions of individual PBs in light and temperature signaling.

This work and the work associated with ARF proteins described above make it clear that subcellular condensates (liquid–liquid phase separation) provide a rich opportunity for applying the PBT approach to understanding how cells use them to segregate and manage populations of biologically active molecules involved in the complex reactions and adaptations of plants to their environment (see also Chapter 8).

Flowering

Another example of plant threshold behavior involves the transition between vegetative and reproductive growth. In general, plants initially grow vegetatively, producing more stems and leaves above ground and more roots below ground until they transition into reproductive growth and produce flowers containing the reproductive organs (ovaries and stamens) for making seeds. Some plants require a period of cold, called vernalization, to make this switch. During vegetative growth, floral development is repressed by the expression of the *FLOWERING LOCUS C* (*FLC*) gene. As duration below a threshold temperature increases, expression of FLC is proportionately inhibited until it is inadequate to prevent the transition to flowering. Interestingly, the cold duration signal continues to accumulate even when it is interrupted by warm periods, suggesting an irreversible and bistable on/off state for *FLC* (Angel et al., 2015). Expression of *FLC* is controlled epigenetically via histone modifications associated with the gene locus (see Chapter 3). "However," as Angel et al. (2015, p. 4146) note, "current measurements of *FLC* histone modification levels are population based and do not reveal the status of individual cells. Consequently, it is unclear whether [histone marker] accumulation during cold temperature exposure is either an all-or-nothing response in each cell (digital) or a gradual increase in all cells (analog)." *In vivo* reporter experiments demonstrated that silencing in response to cold exposure occurs cell-by-cell, and even independently on the alleles of the *FLC* gene on each of the diploid chromosomes (Angel et al., 2011). Angel et al. (2011, p. 108) concluded: "Our combined experimental/modelling approach to vernalization has revealed that the quantitative level of epigenetic silencing in the warm is generated by an appropriate proportion of cells switching into a stable *FLC*-silenced state. Controlled switching of individual cells between bistable states, mediated by nucleation of an opposing histone mark, may therefore be a common method of achieving a quantitative epigenetic response at a population level." This is entirely consistent with the PBT concept that many biological events are based on bimodal individual responses within a population and that either distributions or thresholds can shift in response to inputs such as temperature (Fig. 5.4).

Stomata

Plant leaves contain unique cells that form pores to allow gases (e.g., water vapor, carbon dioxide) to leave or enter the leaf. These cylindrical cells come in pairs and are joined at their ends to form what are called stomata. When these adjacent cells absorb water and swell, their cell walls are constructed in a way that causes them to bulge outward from each other, forming an opening in the center (Fig. 5.5A). By altering the concentration of solutes in these cells, they can either osmotically absorb water from neighboring cells and swell, opening the pore, or they can reduce the amounts of solutes in the cells, losing water and closing the pore. The stomata must open in the light to allow the entry of carbon dioxide for photosynthesis (carbon fixation into stable organic compounds). However, this also allows water to exit the leaf through the same opening. Thus, plants are continuously faced with the economics of

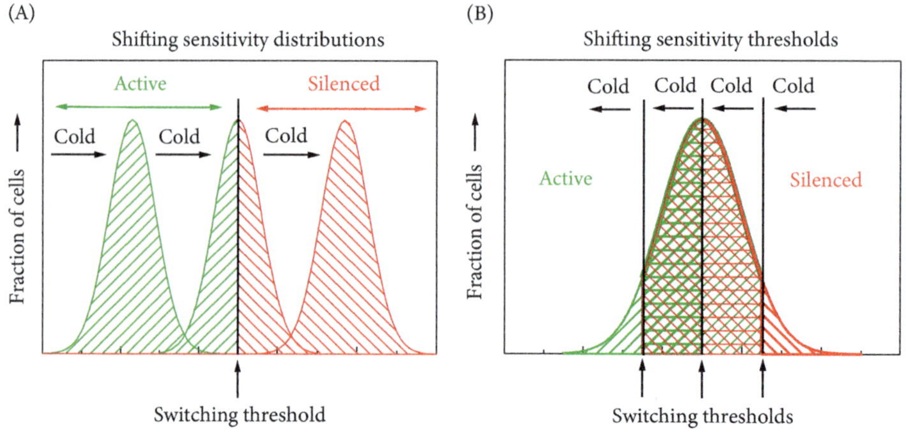

Figure 5.4 Alternative PBT models for switching in a symmetric bistable system, such as *FLC* silencing due to vernalization (cold), enabling flowering (Angel et al., 2011; Angel et al., 2015). **A.** Initially, all cells will be in an active state, expressing the flowering inhibitor FLC. However, exposure to cold temperatures causes the sensitivity distribution among cells to shift toward the threshold at which the *FLC* gene is epigenetically silenced. After extended cold periods, all of the genes will be silenced. **B.** Alternatively, the sensitivities of silencing individual genes could remain the same, but the switching threshold could shift as the cold duration increased. This would have the same effect as shifting the distributions toward a single threshold, as shown in panel **A**, although the cellular/molecular mechanisms might differ between the two alternatives.

spending water in order to gain carbon, controlled primarily by the action of these stomata in opening and closing the pores through which this gas exchange takes place (Farquhar and Sharkey, 1982; Bloom et al., 1985).

As the efficient operation of stomata is critical for successful plant growth, they are subject to complex regulatory interactions (Zhang et al., 2020). Stomata can sense (at least) plant water status, light, carbon dioxide, relative humidity and plant hormones, particularly abscisic acid (ABA). ABA is a plant growth regulator involved in diverse aspects of plant biology, from stomata to seed germination and root growth. However, stomata have become a key single-cell model for exploring the action of ABA (Sirichandra et al., 2009). ABA concentrations increase in response to water stress and cause stomata to close, reducing water loss but also limiting photosynthesis (Bradford and Hsiao, 1982). However, exactly how populations of stomata integrate these diverse signals to adjust their individual responses remains to be fully determined. For example, in response to water stress, "some groups of stomata close while others remain open. The net effect of this response, in contrast to a uniform reduction in stomatal aperture, is a reduction in functional leaf area with proportional decreases in A [carbon assimilation] and stomatal conductance" (Downton et al., 1988, p. 507). ABA is a key signaling molecule indicating water stress, and stomata are more sensitive to ABA after experiencing a prior water stress (Davies, 1978), but it also seems "that individual stomata vary intrinsically in their reaction to ABA" (Terashima et al., 1988, p. 393). In fact, stomatal behavior at the population level is notably variable: "Patchy stomatal conductance shows that the behaviour of things as seemingly elementary as a single leaf can be rather nonlinear and heterogenous, and can defy dynamic prediction" (Mott and Buckley, 2000, p. 261).

This population distribution of response variation among individual stomata is often obscured by using only averaged means to compare the effects of various experimental manipulations (Laisk, 1983). The average pore (aperture) opening clearly shows a log-linear closure response to increasing ABA concentrations (Fig. 5.5A). The black error bars indicate means and standard errors based on 60 individual stomatal aperture measurements. The overall effect of ABA is clear and significant. However, the individual stomatal measurements plotted on a frequency basis show the underlying situation (Fig. 5.5B). In the absence of added ABA, the distribution of apertures among individual stomata is almost perfectly normal, spanning a four-fold range

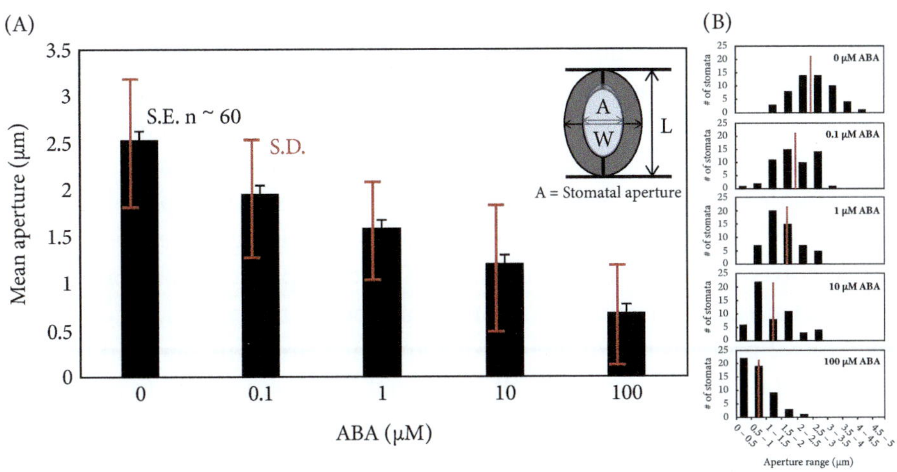

Figure 5.5 Response of stomatal apertures to abscisic acid (ABA) concentrations. **A.** Mean apertures at different ABA concentrations. Black error bars are standard errors ($n = 60$) and red error bars are standard deviations. **B.** Frequencies of stomatal apertures in different size categories at different ABA concentrations. The vertical red lines indicate the mean values.

Adapted with permission (Creative Commons) from Panchal, S., et al. (2016). Regulation of stomatal defense by air relative humidity. *Plant Physiology* 172(3): 2021–2032. https://doi.org/10.1104/pp.16.00696. Individual stomata measurement data in panel **B** provided by Maeli Melotto.

of widths from nearly closed to maximally open (Fig. 5.5B). As the ABA concentration is increased (logarithmically), the mean of the distribution shifts to lower apertures, as was indicated by the mean values (Fig. 5.5A). However, the entire distribution shifts in concert as the ABA concentration increases, maintaining essentially the same standard deviation (red error bars in Fig. 5.5A). Even at 100 μM ABA, some stomata are still almost as open as the mean in the absence of ABA. This population-based behavior of stomata was recognized and modeled decades ago (Laisk et al., 1980; Laisk, 1983) but has received relatively less attention from those studying stomatal behavior. This behavior clearly illustrates the basic thesis of the PBT model that individuals can vary in their sensitivities to regulatory factors, but collectively, their behavior can conform to what might be expected of the average. However, this does not mean that differences in their behavior across four orders of magnitude of hormonal concentrations is simply experimental error or "noise." Rather, it indicates that standard errors, due to their sensitivity to sample number, often obscure the size of natural variation that is present. In this case, 60 individual measurements per ABA concentration resulted in small standard errors and clear mean differences, despite the extensive overlaps among the total populations. In fact, essentially any two nonidentical means can be demonstrated to be significantly different if enough data points are collected (Clayton, 2021).

The sensitivity variation among stomata has now been explicitly demonstrated by monitoring the responses of individual stomatal cells to ABA. SnRK2 (sucrose-non-fermenting-1-related protein kinase-2) proteins are involved in ABA signal transduction in plants (Nakashima et al., 2009; Umezawa et al., 2009). Perception of ABA activates SnRK2 proteins which phosphorylate transcription factors and other physiological regulators associated with stresses, seed germination and other responses (Nonogaki, 2020). Zhang et al. (2020) developed a fluorescent reporter that would indicate when the SnRK2 proteins responded to ABA and measured this signal in individual stomatal cells. Even though all epidermal cells were affixed to a slide and bathed in a solution of the same ABA concentration, the magnitude and speed of response varied widely among individual stomata (Fig. 5.6A). This type of variation is often referred to as "noise" and is analyzed by statistically identifying only a mean response and standard errors. In contrast, using the PBT model, I was able to decompose the raw data into at least three response subgroups (Fig. 5.6B, C). One group showed a rapid and sustained response (Fig. 5.6B, red curves), another had a strongly delayed and reduced maximum response (Fig. 5.6B, blue curves) and yet another exhibited a rapid initial response followed by a linear decline over time (Fig. 5.6C). In each

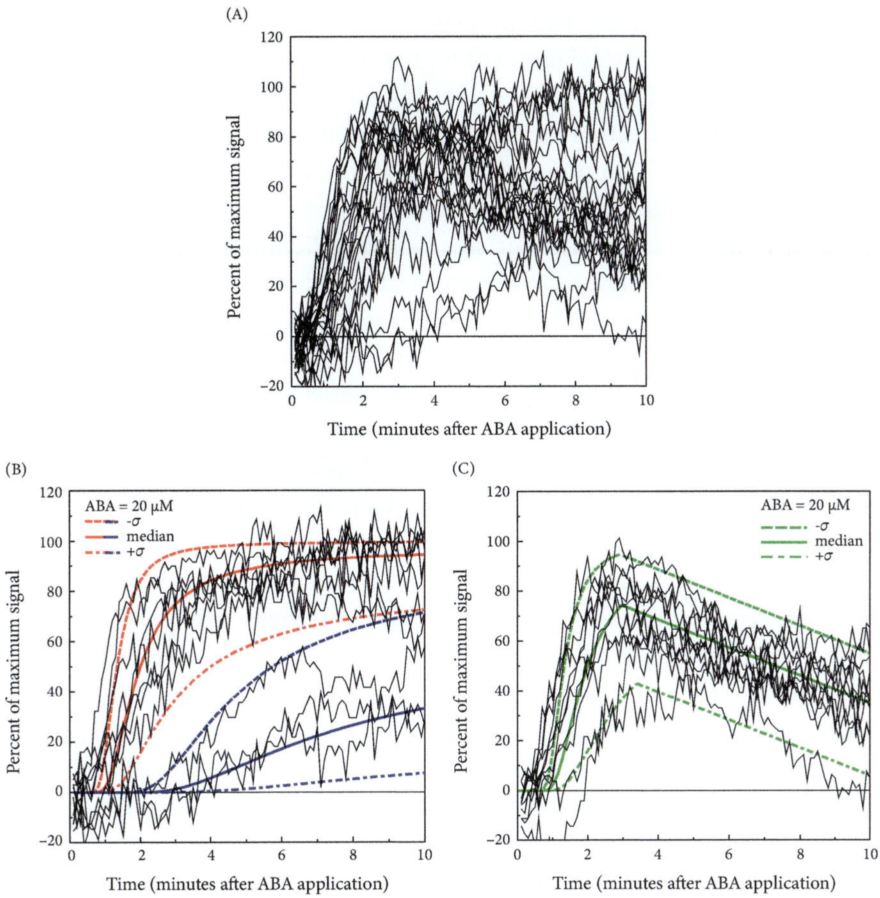

Figure 5.6 Single stomatal cell monitoring of SnRK2 activation by abscisic acid (ABA). **A.** Raw data with each line indicating the response of an individual stomate to exposure to 20 μM ABA at time 0, normalized to the maximum response observed. **B.** Subfractions of cells that showed a continuing (or asymptotic) maximum response to ABA. These separated into two categories (red and blue lines) based on time of initiation and maximum signal intensity of the responses. Predicted responses to 20 μM ABA based on the PBT model (see Fig. 5.7) are shown for the median response and one standard deviation (σ_{ABA}) below and above the median for comparison to the observed range of responses **C.** Remaining stomata exhibited an initial response predicted by the PBT model followed by a linear decline in signal output (green lines). PBT model parameters for the initial response were the same as those for the red subpopulation in panel **B.**

Raw data reproduced with permission (Creative Commons) from Zhang, L., et al. (2020). FRET kinase sensor development reveals SnRK2/OST1 activation by ABA but not by MeJA and high CO_2 during stomatal closure. *eLife* 9: e56351. https://doi.org/10.7554/eLife.56351.

case, an estimation of the median response is shown along with estimates for shifts of one standard deviation (σ) higher or lower on the sensitivity distributions (shown in Fig. 5.7). There could be functional significance to these subpopulations, indicating inherent variation in sensitivity and in physiological responses to ABA among individual stomata. In addition, unpublished data provided by the authors showed more rapid responses at higher ABA concentrations (not shown), consistent with predictions of the PBT model. Of course, this raises the issue of what causes the differential sensitivity of stomatal subpopulations to ABA concentrations. As noted previously for other sensors, the simplest explanation could be that the number of SnRK2 proteins might vary among individual stomata. Alternatively, post-translational modifications could alter their individual sensitivities to ABA, such as in response to environmental signals. Relaxation or feedback from other response pathways might be associated with the linear decrease in signal output with time in some stomata (Fig. 5.6C).

The ability to visualize the activity of individual sensor molecules would allow experimental tests of these hypotheses, perhaps using methods similar to the imaging and *in silico* methods developed to synchronize the cell division times of a cell population (Cappell et al., 2018). Stomatal behavior is known to be "patchy," or spatially variable, as it is influenced by multiple local environmental inputs as well as hydraulic and mechanical aspects of the pressure-driven opening mechanism (Mott and Buckley, 2000). Models attempting to reproduce this behavior have needed to introduce random spatial variability among individual stomata (Haefner et al., 1997). Thus, the "patchiness" in stomatal behavior could be the result of the combination of inherent sensitivity differences among the stomata and the unique microscale environments that they are sensing at their location on the leaf. This includes considerable cell-to-cell and tissue variation in endogenous ABA levels, as revealed by *in vivo* fluorescence (Forster resonance energy transfer, or FRET) sensors (Rowe et al., 2023). In addition to controlling water loss, in which ABA is a primary signal, stomata also monitor CO_2 levels, humidity, light and other environmental conditions to optimize the economics of carbon gain and water loss, consistent with the instantaneous conditions they are experiencing (Fan et al., 2025). Thus, we can imagine that an individual stomate would also have sensitivity thresholds for CO_2, humidity, light and so on that could vary from its neighbors. In fact, a stomate would likely be a member of multiple sensitivity subpopulations for different factors, and the result for gas exchange at the leaf level would be the net effect of these collective responses across the stomatal population to the conditions they are experiencing at their specific location. Once these distributions were

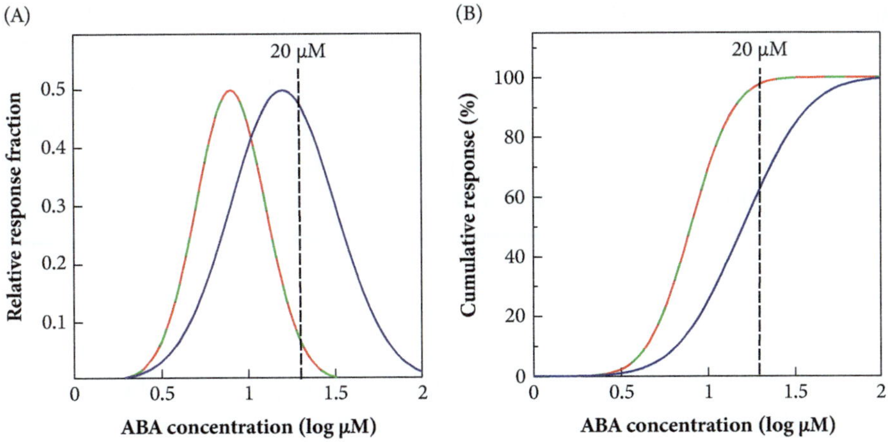

Figure 5.7 Estimated response distributions for stomatal responses to ABA using the PBT model based upon data in Figure 5.6. **A.** Estimated sensitivity threshold distributions for individual stomata to respond to ABA in the red, green and blue subpopulations shown in Figure 5.6B, C. The red and green subpopulations have the same sensitivity distribution. The concentration used in the experiment (20 μM) is indicated. Note that the ABA concentration axis is on a logarithmic scale. PBT model parameters were θ_{ABA} = 0.8 log μM min, $X_{ABAb}(50)$ = 0.9 log μM, and σ_{ABA} = 0.2 log μM for the red and green subpopulations and θ_{ABA} = 2.3 log μM min, $X_{ABAb}(50)$ = 1.2 log μM, and σ_{ABA} = 0.3 log μM for the blue subpopulation. **B.** Cumulative percentage of responding stomata as ABA concentration increases logarithmically. The model indicates that the most sensitive stomata in the red and green subpopulations would respond at about 3 μM ABA, while all stomata in the blue subpopulation would not respond until the ABA concentration exceeded 60 μM.

established, regulation of overall leaf conductance would be automatic as the levels of those input factors varied; the PBT model can accommodate summing of simultaneous responses to multiple parameters (Ni and Bradford, 1992, 1993; Bradford and Bello, 2022). In addition, these sensitivities and time constants need not be fixed, but they could vary diurnally or in response to stress, with cells being able to turn their sensitivities up or down by activating or deactivating their sensor molecules or by changes in the activity of enzymes synthesizing or degrading hormones. Seller and Schroeder (2023) have demonstrated dynamic chromatin remodeling in response to ABA that was associated with cell-type specificity. Light, for example, could be an overriding signal that might need to be satisfied before the other pathways could be active, as stomata remain closed in the dark. The development of methods allowing individual stomata to be observed and their responses to be quantified (Zhang et al., 2020; Rowe et al., 2023) would allow this hypothesis to be explored by conducting dose-response studies to different input (or internal) factor levels at a specific spatial location on a leaf and quantifying the responses of populations of individual stomata. This could be combined with spatial single-cell transcriptional profiling to associate specific responses with, for example, expression of receptor proteins associated with sensitivity thresholds (Liu, Z. et al., 2020; Xia et al., 2022).

Seeds

Plant seeds are the paradigmatic individuals whose population behavior was the first described by the PBT approach (Gummerson, 1986; Bradford, 1990). While there are philosophical issues with the concept of what "individual" means in biology, due to the multiple levels of organization involved in a single organism and their interconnectedness (Bueno et al., 2018), there can be little argument that seeds are biological individuals. In nature, the purpose of seeds is to disperse as individuals to spread their genetic content broadly, in both space and time. Each seed truly is on its own to sense its current environment and physiological status and to make the irrevocable decision regarding whether to germinate or not. We are perhaps misled in this matter because of our experience with seeds of domesticated crop plants, which have been intensely selected to germinate immediately when given water at an adequate temperature. Thousands of years of selection and domestication by agriculturalists have eliminated most of the dormancy mechanisms present in wild seeds to time their germination and emergence to coincide with favorable times of the year and specific environmental conditions (particularly water, temperature and light). Native seeds have remarkably complex systems of both dispersal and dormancy that serve to spread them widely from their parent plant and enable them to remain dormant until specific environmental conditions are met (Baskin and Baskin, 2014, 2021). Some examples from seed biology relevant to the PBT model are given below; the capabilities of seeds in an ecological sense will be addressed in Chapter 7. A more complete discussion of seed biology and PBT models can be found in Bradford and Bello (2022).

Seed dormancy and longevity

A discussion of the population-based nature of seed biology must start with recognition of the landmark vision and research of Eric Roberts. Roberts was the first to recognize that the loss of dormancy and eventually of viability of rice (and other cereal) seeds during dry storage occurred in patterns fitting a cumulative normal distribution (Roberts, 1960, 1961). That is, a population of dormant rice seeds would increase in germination capacity over time in a sigmoid manner, exactly representing a normal distribution of the times of dry storage required to alleviate dormancy among seeds. Similarly, seeds stored for longer times (or under more adverse conditions, such as higher water content or temperature) exhibited a declining sigmoid curve representing the loss of viability among the individual seeds. With his colleague Richard Ellis, Roberts developed population-based equations representing the relationships among temperature, water content (or equilibrium relative humidity) and longevity that have been widely adopted in seed storage and conservation (Ellis and Roberts, 1980, 1981; Roberts and Ellis, 1989; Ellis, 2022). This approach was subsequently extended via the PBT model to also describe the increases in germination speed and percentage as seed dormancy was lost during dry storage (Alvarado and Bradford, 2005) and as the seeds required an increasing amount of time to germinate as they aged before losing viability (Bradford et al., 1993), apparently due to the increasing time required for repair mechanisms to operate after seed hydration (Reed et al., 2022; Waterworth and West, 2023). The PBT model fits these patterns exceptionally well, and they serve as clear illustrations of the population basis of seed germination and longevity patterns (Bradford and Bello, 2022).

Temperature and water potential

Seed germination has long been known to be sensitive to temperature and to proceed on a thermal time basis (Bierhuizen and Wagenvoort, 1974). That is, the temperature [T] above a base temperature [T_b] multiplied by the time to germination [t_g], or $(T\text{-}T_b)\ t_g$, is a constant, representing the thermal time (degree-days) required for a given seed to germinate after initial hydration (imbibition). Thermal time is used across plants (and insects) in agriculture to convert variable weather conditions into cumulative degree-days, which more accurately predict their development and maturation than calendar days (Trudgill et al., 2005). Gummerson (1986) subsequently showed that a similar approach could be applied to germination responses to water limitation or water potential (related to chemical activity). However, unlike base temperature, which is often a common value within a seed population, Gummerson found that seeds varied in a normal distribution in their minimum water potentials permitting germination. Combining temperature and water potential into a single equation, he developed the "hydrothermal time" model that was the basis for the PBT model (Chapter 2; Bradford, 1990; Alvarado and Bradford, 2002). As this hydrothermal time model is covered in detail elsewhere (Bradford and Bello, 2022), I will simply note that a literature search using the terms *hydrothermal time* and *seed* identified more than 300 relevant publications, indicating the broad adoption and application of these concepts across seed biology.

Hormonal sensitivity of germination

From a physiological point of view, seeds exhibit multiple population-based behaviors that illustrate the generality of the PBT concept. This can be demonstrated by the responses of seed germination to a hormonal promoter (gibberellin, GA) or inhibitor (abscisic acid, ABA) (Ni and Bradford, 1993). In this case, we used tomato seeds from inbred genetic lines that had mutations preventing the seeds from making their own supplies of these hormones; that is, they were either GA-deficient (*gib-1*) or ABA-deficient (sit^{w}). Since GA is an essential requirement for germination, the *gib-1* seeds were completely dependent upon externally supplied GA in order to germinate (Fig. 5.8A). ABA, on the other hand, is a strong endogenous inhibitor of germination, so sit^{w} seeds germinated more rapidly and to a higher percentage than did the wild-type seeds without the mutation, but they were inhibited by the addition of ABA (Fig. 5.8B).

The ranges of variation in sensitivity thresholds among seeds illustrate the two major components of the PBT model: recruitment and kinetics. It is evident that at intermediate hormone concentrations, the final germination percentage reaches a plateau (asymptotic) value well below 100%. This immediately indicates that a given hormone concentration is effective for only a specific fraction of the population and that larger fractions of the population respond (positively or negatively) as the concentrations increase (Fig. 5.8E, F). It also illustrates the recruitment component, as shown by the changes in final germination percentages with increasing GA concentration (Fig. 5.8E). In addition, the germination of those newly recruited individuals occurs more slowly compared to seeds sensitive to lower concentrations. That is, as GA concentrations increase, all seeds germinate more rapidly, in direct proportion to the amount by which the concentration exceeds their individual thresholds. Thus, there is a larger relative kinetic effect for the less sensitive seeds compared to the more sensitive seeds as GA concentration increases across the sensitivity distribution. This results in the

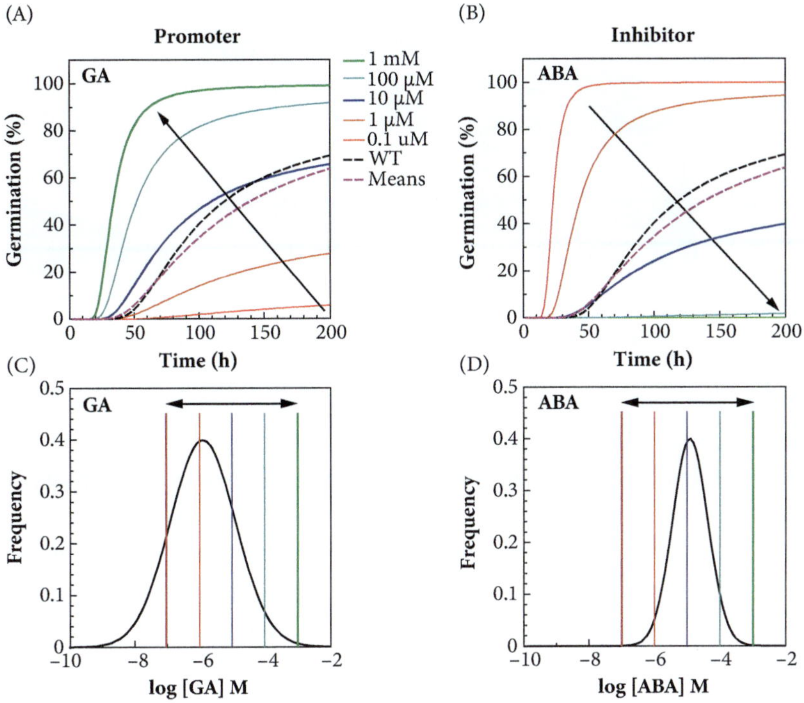

Figure 5.8 Modeling the effects of gibberellin (GA) and abscisic acid (ABA) on germination timing and percentage of tomato seeds. **A.** Germination time courses predicted for GA-deficient (*gib-1*) tomato seeds imbibed in various concentrations of GA. These seeds cannot germinate without added GA, and they increase in germination percentage and rate (or speed, i.e., the time to germination decreases) as the concentration increases. **B.** Germination time courses predicted for ABA-deficient (*sit*^{W}) tomato seeds imbibed in various concentrations of ABA. These seeds germinate rapidly and completely without added ABA, and both germination percentage and rate (speed) decrease as the concentration increases. For panels **A** and **B**, the fitted curves matched the actual data with r^2 values of 0.97 and 0.91, respectively (Ni and Bradford, 1993). Also, in panels **A** and **B**, the black dashed line represents the germination pattern of wild-type seeds that manage their own internal GA/ABA balance. The purple dashed line indicates what the PBT model would predict for wild-type seeds if their internal hormone concentrations were effectively the same as the means of the respective sensitivity distributions. **C.** Sensitivity threshold distribution for GA-deficient (*gib-1*) tomato seeds. Vertical lines indicate where the GA concentrations in panel **A** occur in relation to this distribution. **D.** Sensitivity threshold distribution for ABA-deficient (*sit*^{W}) tomato seeds. Vertical lines indicate where the ABA concentrations in panel **B** occur in relation to this distribution. **E.** Germination time courses from panel **A** shaded to indicate that both recruitment (increasing percentage) and kinetics (more rapid germination) of *gib-1* seed germination are affected by increasing GA concentrations. **F.** Germination time courses from panel **B** shaded to indicate that both recruitment (decreasing percentage, as ABA is an inhibitor) and kinetics (more delayed germination) of *sit*^{W} seed germination are affected by increasing ABA concentrations.

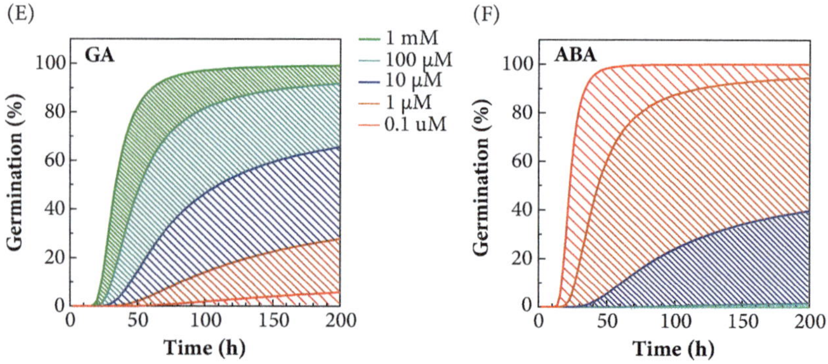

Figure 5.8 Continued

skewness of the time courses, even though the underlying sensitivity distribution is symmetric (on a log scale) (Fig. 5.8C), as is evident in the defining equation of the PBT model. In this model, for a given time constant value (θ_X), the time to germination (t_i) will be inversely proportional to the difference between the minimum (or base) threshold of that fraction and the current concentration:

$$t_i = \theta_X/[X - X_b(i)] \tag{5.1}$$

where *i* refers to the fraction of the population, *X* is the factor (GA concentration in this case) and $X_b(i)$ is the distribution of thresholds for factor *X* in the population. The same logic applies to the inhibitory effect of ABA, in which germination speed and percentage are maximal at low concentrations, and germination is progressively delayed and prevented as the ABA concentration increases across the sensitivity distribution (Fig. 5.8B, D, F). This inhibitory effect can be expressed in the model simply by reversing the difference term:

$$t_i = \theta_X/[X_b(i) - X] \tag{5.2}$$

The wild-type seeds (which adjusted their own internal GA/ABA balance) exhibited some dormancy, as only about 70% of seeds completed germination in water (the black dashed line in Fig. 5.8A,B). If the sensitivity distributions in the wild-type seeds for GA and ABA were the same as in the *gib-1* and *sit*^{w} mutants (which is not necessarily the case), then we might ask what would happen if we modeled the effect on germination of providing GA and ABA concentrations simultaneously at the mean sensitivity to each hormone (i.e., 1.26 µM for GA and 12.6 µM for ABA)? The predicted germination time course is shown by the purple dashed lines in Figs. 5.8A,B. While not perfect (the fit can be improved by minor adjustments to the parameters), it is striking how this prediction almost matches the actual behavior of the wild-type seeds (black dashed lines) in the absence of added hormones. While certainly not hard evidence, this suggests that the wild-type seeds might be functioning with effective endogenous GA and ABA concentrations near to the medians of their respective sensitivity distributions. Of course, this is in relation to the effect of added exogenous hormones, which do not have uniform effects across the cell populations in seeds (Topham et al., 2017; see Fig. 5.10), and endogenous hormonal concentrations need not be uniform across tissues (Herud-Sikimić et al., 2021; Drapek et al., 2024; Griffiths et al., 2024). Nonetheless, this simple exercise indicates that we can quantify at least the relative influences of these two hormones *in vivo* in the cells controlling germination.

Enzyme activities

A prediction of the PBT approach to understanding physiological variation is that we would expect the individuals in such populations to vary in measurable phenotypes consistent with their position in the population distribution. For the experimental system described above, the act of germination (i.e., the emergence of the embryonic radicle [root] from the enclosing endosperm and seed coat) for an individual seed should be associated with the increased activity of an enzyme that can degrade cell wall components in the enclosing endosperm, thereby weakening this tissue and allowing the embryonic radicle to emerge. Thus, we measured the activity of endo-β-mannanase (a cell wall-degrading enzyme) in the relevant tissues (the endosperm cap enclosing the embryonic radicle) of individual wild-type tomato seeds at different times after hydration (Still and Bradford, 1997). What is remarkable about the results is that after 24 hours of hydration, prior to any radicle emergence, endo-β-mannanase enzyme activity varied over 3000-fold, from 0.1 pmol min^{-1} to 300 pmol min^{-1}, among different individual seeds in the population (Fig. 5.9A). After 48 h, just as some seeds were beginning to germinate, the population had diverged into two distinct subpopulations, one with activities of 10 pmol min^{-1} or less and the other with activities between 10 and 1000 pmol min^{-1} (Fig. 5.9B). At longer times, as additional seeds germinated (and were no longer measured), the remaining seeds in the higher-activity population disappeared, while the low-activity subpopulation persisted and was associated with the dormant seeds remaining ungerminated at the end of the time course (Fig. 5.9E). This result indicates that seeds in the subfraction with very low endo-β-mannanase activity represented the dormant seeds that would not complete germination, while seeds that highly expressed the enzyme eventually germinated. In addition, providing 100 μM GA to wild-type seeds eliminated the low-activity subfraction of the bimodal distribution of enzyme activity and all the seeds germinated. Also consistent with the germination patterns in Fig. 5.8, GA-deficient *gib-1* seeds exhibited an increased range of activities across a logarithmic scale when provided with 10 μM GA, which results in a germination time course similar to that in wild-type seeds (Fig. 5.8A). ABA-deficient sit^{w} seeds, on the other hand, exhibited relatively higher enzyme activities, including some at 10,000 pmol min^{-1}. In fact, "because of the logarithmic range in mannanase activity, only three seeds of 45 with exceptionally high activity accounted for 93% of the total observed at this sampling point and shifted the mean 80-fold higher than the median" (Still and Bradford, 1997, p. 28). This indicates the potential scale of error (and misinterpretation) that can occur when all individuals in a diverse population are pooled together for sampling.

Signal perception and response

The interaction between GA and ABA in regulating seed germination brings up an additional issue: what are the targets of hormone action in seeds? As described above, specific tissues (e.g., the endosperm) can respond to hormones differently from the embryo (Nonogaki et al., 2000, 2007). Topham et al. (2017) took this a step further and showed that the mutually inhibitory interactions of GA and ABA can be modeled as a bistable switch underlying the incorporation of temperature variations into germination decisions. They also showed that the ABA and GA response centers are housed in distinct cellular regions in the Arabidopsis embryo (Fig. 5.10A). In fact (p. 6629), "this spatial separation within the decision-making center is required to process variable temperature inputs from the environment to promote the breaking of dormancy. In contrast to other noise-filtering systems, including human neurons, the functional role of this spatial embedding is to leverage variability in temperature to transduce a fate-switching signal within this biological system." In this sense, it is like the Turing system mentioned above, in that signal movement across a distance is integral to the regulatory process. They subsequently used "stochastic chemical kinetic modelling to analyse a hormone feedback signalling motif in *Arabidopsis thaliana* seeds that can generate tunable levels of noise in the hormone ABA,

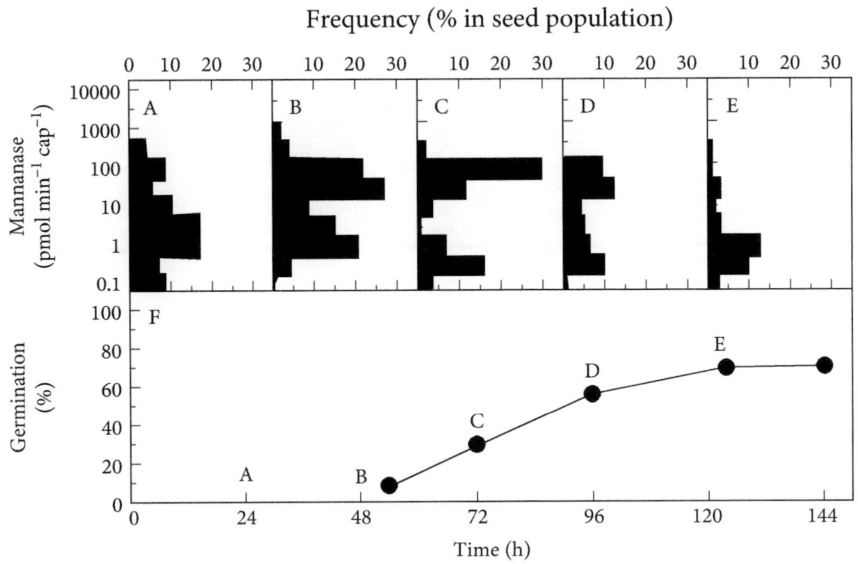

Figure 5.9 Frequency distributions of endo-β-mannanase activity in individual endosperm caps excised from tomato seeds at different time points after imbibition. Panels **A, B, C, D** and **E** show frequencies of endo-β-mannanase activity (on a log scale) in remaining ungerminated seeds at each time, as indicated in panel **F** along with the cumulative germination percentage (based on the total number of seeds tested). Only dormant seeds remained at the time point in panel **E**.

Reproduced with permission from Still, D.W. and Bradford, K.J. (1997). Endo-β-mannanase activity from individual tomato endosperm caps and radicle tips in relation to germination rates. *Plant Physiology* 113(1): 21–29. https://doi.org./10.1104/pp.113.1.21.

governing germination propensity. The key feature of the motif is simultaneous positive feedback regulation of both ABA production and degradation pathways, allowing tunable noise while retaining a constant mean level" (Johnston and Bassel, 2018, p. 1). Since we can treat seeds in bulk with hormone solutions and modify their behavior, we have unconsciously assumed that the hormonal concentrations, and therefore the cellular responses, occur generally in all the embryo cells. However, the *in vivo* responses to ABA and GA are localized to relatively few cells near the radicle tip, while the growth response required for germination does not occur in that region, but farther back in the hypocotyl region (Sliwinska et al., 2009) (Fig. 5.10B).

While this work demonstrates the differential sensitivities of embryo cells to ABA and GA, it is also the case that these sensitivities and their associated times to germination vary among individual seeds (Ni and Bradford, 1993). In fact, Abley et al. (2021) found that times to germination even varied among individual seeds from the same fruit (silique) in Arabidopsis.

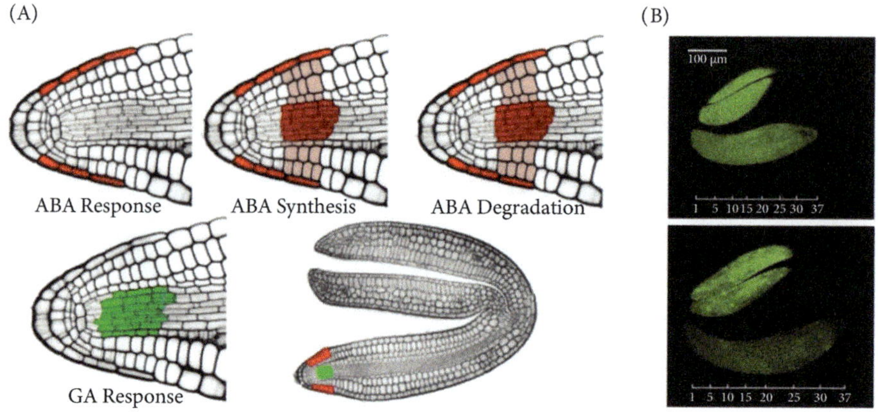

Figure 5.10 Hormonal signal perception and response in plant embryos. **A.** Locations of ABA response, synthesis and degradation and of GA response in *Arabidopsis thaliana* embryos. As the drawing of the entire embryo shows, the cells responding to ABA (red) and to GA (green) are localized to distinct areas in the radicle tip. **B.** While hormonal signals are received and processed at the radicle tip, the actual cellular growth that drives embryo expansion and germination is located distal from the radicle tip. The top panel shows an embryo 22 h after imbibition, before expansion has begun, and the lower panel shows an embryo 50 h after imbibition. The scale at the bottom is cell number from the point of attachment of the cotyledons to the radicle tip. The cells between 20 and 30 have visibly expanded, while those near the tip (between 30 and 37) have not.

Panel A reproduced with permission from Topham, A.T., et al. (2017). Temperature variability is integrated by a spatially embedded decision-making center to break dormancy in *Arabidopsis* seeds. *Proceedings of the National Academy of Sciences USA* 114(25) 6629–6634. https://doi.org/10.1073/pnas.1704745114. Panel B reproduced by permission from Sliwinska, E., et al. (2009). Germination of *Arabidopsis thaliana* seeds is not completed as a result of elongation of the radicle but of the adjacent transition zone and lower hypocotyl. *Journal of Experimental Botany* 60(12): 3587–3594. https://doi.org/10.1093/jxb/erp203.

This result suggested that it is an inherent feature of seed development and maturation, rather than being due to genetic or environmental variation (see also Chapter 7). Through genetic analyses, Abbey et al. concluded that this variability could be due to differences in the sensitivities of individual seeds to ABA. They developed a model based on a double positive-feedback loop between ABA and GA that can vary the ABA sensitivity and included a stochastic noise generator to account for variation among seeds in germination timing. This model also resulted in a bistable switch and could describe the effects of exogenous ABA and GA on germination and its variation. While Abbey et al. did not specifically apply or test the PBT model, their model indicates that molecular mechanisms are present that can satisfy the underlying expectations and relationships with respect to the effects of these hormones on germination, that is, variations in sensitivity among seeds and responses in both the percentages and timing of germination. As the authors note (p. 24), "to fully understand the roles of germination regulators, future work should involve characterising their effects on the CV [coefficient of variation] of germination time distributions as well as on traits such as percentage germination and average germination time." These are exactly the components that are specifically described by the PBT model.

Seed-to-seed variation

Pursuing the issue of seed-to-seed variability, Krzyszton et al. (2022) extracted and analyzed transcriptomes from genetically identical individual Arabidopsis seeds as they were induced into secondary dormancy by exposure to high temperature after imbibition. They observed that transcriptional variation decreased as dormancy was induced and increased again after treatments to break dormancy. Cao (2022, p. 1561) interpreted these results in a manner consistent with the PBT model in which the transition into dormancy occurs on a seed-by-seed basis (Fig. 5.11):

> The findings presented by Krzyszton et al. (2022) build a graceful bridge between the germination competence of single seeds and germination behaviors of seed populations at the transcriptional level. The authors showed that stress treatments gradually induced a decrease in inter-seed transcriptional heterogeneity, along with directional changes in expression patterns of key genes associated with dormancy regulation [Fig. 5.11]. The transcriptional germination competence of each seed determines its binary outcome of germination or not. The proportional distribution of binary outcome of all seeds in a population determines the germination percentage.

Krzyszton et al. (2024) subsequently demonstrated that inherent variation in individual seed sizes was associated with expression of a gene promoting seed dormancy, *DELAY OF GERMINATION1* (*DOG1*) (Bentsink et al., 2006). Smaller seeds tended to be more dormant and have higher expression of *DOG1* compared to larger seeds. An extreme case of such seed dormancy variation has been demonstrated in cocklebur (*Xanthium strumarium*), which produce both a smaller, more dormant seed and a larger, less dormant seed in each individual fruit (Esashi and Leopold, 1968). Nemati et al. (2022) demonstrated that the smaller, dormant seeds had higher

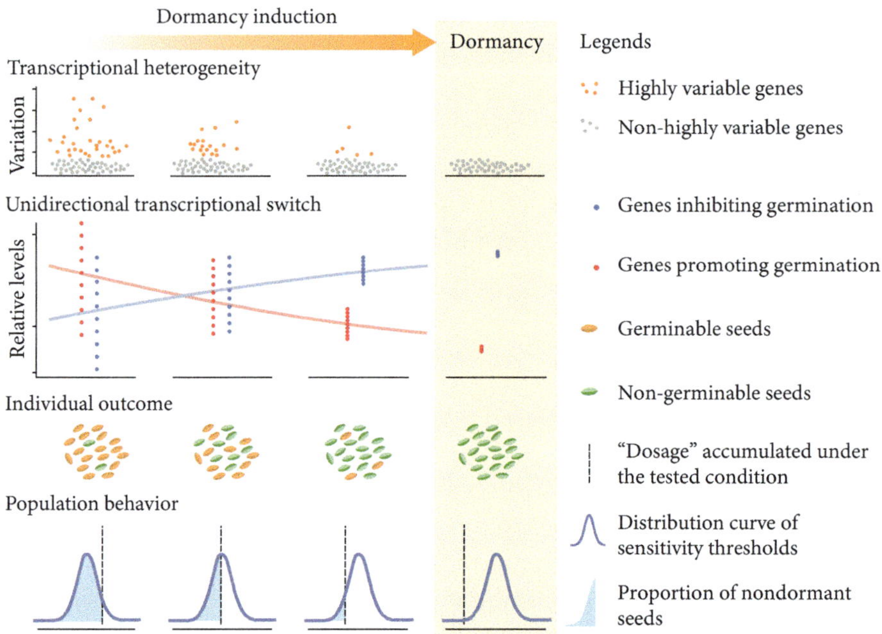

Figure 5.11 The variation in transcriptomes among individual seeds (and genes) decreases as dormancy is induced in the seed population. These transcriptomic switches in each seed determine whether it will or will not germinate; that is, transcripts of genes associated with germination decrease while those associated with inhibition of germination increase. In the dormant state in which metabolic activity is minimal, there is low variability among seeds. This seed-by-seed transition is illustrated as the change from orange (nondormant) to green (dormant) seeds as an increasing fraction of seeds in the population are switched into the dormant state.

Adapted with permission from Cao, D. (2022). To germinate or not? Transcriptional gradients underlie the seed dormancy continuum. *Plant Physiology* 190(3): 1559–1561. https://doi.org/ 10.1093/plphys/kiac367, describing work by Krzyszton et al. (2022).

expression of *DOG1* than did the larger, nondormant seeds. *DOG1* action also interacts with environmental conditions (e.g., temperature) that affect plant life cycles (Vidigal et al., 2016) and with miRNAs involved in the regulation of flowering time (Huo et al., 2016). Thus, variation in seed size and dormancy can be linked to reproductive adaptations to local environments and evolution (Auge et al., 2019; see also Chapter 7).

Cell-to-cell variation within seeds

As in other fields of biology, seeds have also been examined at the single-cell or single-nucleus level, allowing distinct cell types that develop during embryogenesis to be identified by their unique transcriptional complements and spatial patterns. Pelletier et al. (2025) examined 41 subregions of developing soybean (*Glycine max*) seeds and noted (p. 1) that "[t]he number of cell identities exceeds the number of anatomically distinguishable cell types, emphasizing the spatial complexity of seeds." Liew et al. (2024, p. 1429) followed the single-cell transcriptional states of Arabidopsis seeds during their transition to germination and concluded that "the cells of the embryo progress through specific transcriptional states as germination progresses. This enables individual cell types to express the genes that define the changing functions of those cell types at the appropriate time, thereby contributing to the successful seed–seedling transition." The individuality of seeds extends down to their cellular level, with much more to be learned about how transition from quiescence to germination is controlled and effected.

Subpopulations

Seeds also offer an excellent opportunity to illustrate the consequences of the presence of subpopulations within a larger population. It is not unusual in seed germination testing to find seed lots that do not show a single sigmoid time course, but rather one exhibiting skewed or multiple components. Since each seed can germinate only once, the occurrence of such time courses indicates that in those cases, the seeds do not all constitute a common distribution (Bello and Bradford, 2016; Bradford and Bello, 2022). Multiple reasons explain the presence of subpopulations in a seed lot, such as the environment during development, maturity at harvest, dormancy, aging, purposeful blending of commercial seed lots and numerous other factors (Abley et al., 2016; Auge et al., 2019; Fernández Farnocchia et al., 2021; Bradford and Bello, 2023). Fortunately, it is easy to expand PBT models to account for subpopulatons by summing two (or more) separate models and introducing a factor multiplying each model equation by the fraction of the total population represented by that subdistribution (see the Appendix) (Bello and Bradford, 2016; Bradford and Bello, 2022). Figure 5.12 presents several common situations in which subpopulations could be present. For example, if two separate seed lots with different sensitivity threshold distributions were combined in equal amounts (Fig. 5.12B), the effects would be visible in the resulting germination time course of the combined population. Moreover, fitting two models simultaneously to the time course data can determine the percentages of each subpopulation in the lot and the PBT model parameters of each (Bello, 2020). Mixtures with unequal fractions of each subpopulation also alter the time courses in distinctive ways (Fig. 5.12C, D). Changing

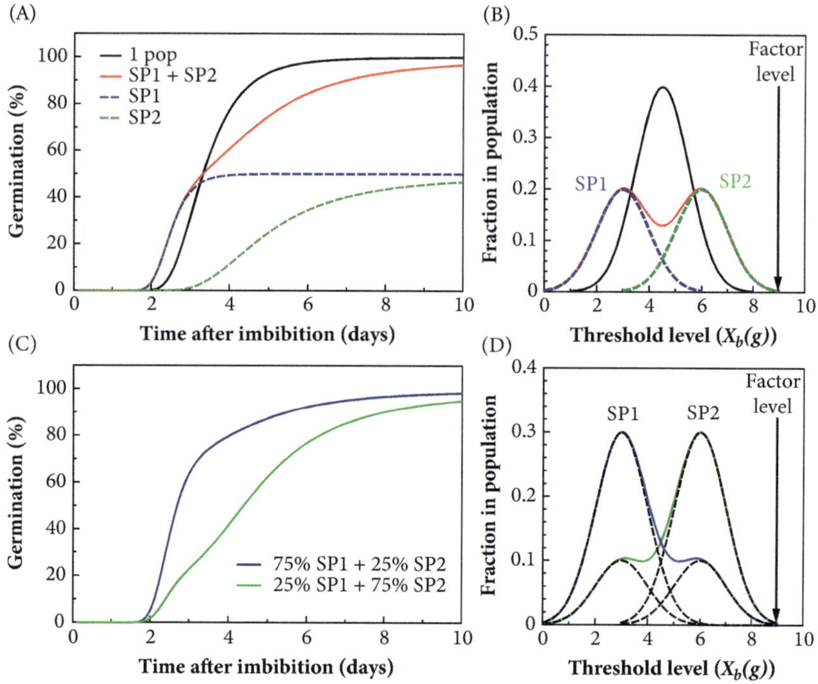

Figure 5.12 Illustration of the flexibility of the PBT model when extended to include multiple subpopulations. **A.** Germination time courses based upon a single population model (black line) or upon two distinct subpopulations blended together in equal numbers (red line). The two subpopulations each have different germination kinetics, as shown by the blue and green dashed lines. **B.** The sensitivity threshold distributions of subpopulation 1 (SP1), subpopulation 2 (SP2) and a single population using the average mean and sigma values of SP1 and SP2. These distributions are the basis of the germination time courses in panel **A,** using a common time constant (θ_X) value of 15. The factor level is indicated for a germination-promoting factor *X* (e.g., Fig. 5.8A). The red line indicates the sum of seeds germinating in both SP1 and SP2. **C.** Effect of different proportions of SP1 and SP2 on germination time courses. The germination time courses indicate those expected of a 75% SP1 + 25% SP2 mixture (blue line) or a 25% SP1 + 75% SP2 mixture (green line). **D.** Illustration of different fractions of SP1 and SP2 (with the same threshold distributions as in panel **B**). The blue and green lines indicate the fractions of each SP in the total and the same time constant as in panel **A**. **E.** Effects of changing σ (standard deviation) and θ_X on germination time courses. The σ values of both SP1 and SP2 were reduced by half (from 1 to 0.5) and seeds from both were blended in equal quantities (blue line). The purple and orange lines indicate the effects of different θ_X values but the same sensitivity distributions on germination time courses. This illustrates that the time constant can either exacerbate differences (purple) or compensate for them to a result in a time course mimicking a single population (orange). **F.** The sensitivity distributions of SP1 and SP2 when their standard deviations are reduced from 1 to 0.5. Time courses in panel E are based on these distributions using values of θ_X of 15 for both (blue), 10 for SP1 and 20 for SP2 (purple line) and 20 for SP1 and 10 for SP2 (orange line).

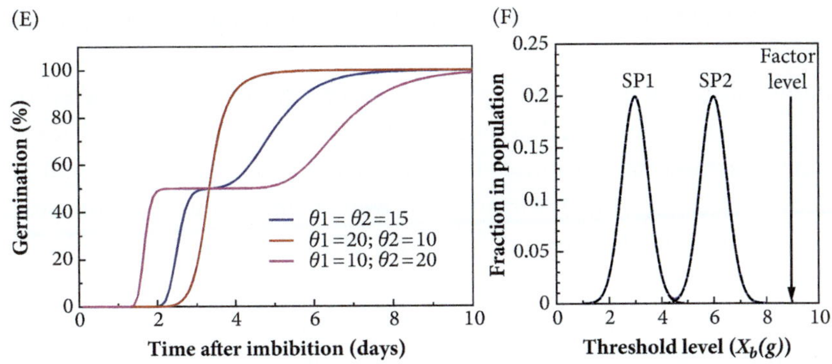

Figure 5.12 Continued

the range of variation (e.g., increasing or reducing their standard deviations) can either emphasize such differences or make them less evident (Fig. 5.12E), as can changing their mean values (not shown). Each distribution can also have distinct time constants, which can also enhance or hide the underlying variation (Fig. 5.12E). The factor level is the same in these examples; if the subpopulation distributions differ, their kinetic and percentage responses to a regulatory factor will also differ, resulting in additional time-course variation. In addition, seeds respond to multiple regulatory factors, so subpopulations could also be differentiated due to differences in their sensitivities to those factors, such as to ABA and GA (Fig. 5.8; Ni and Bradford, 1993). The overall point is that by varying the sensitivity distributions, time constants and fractional contributions among subpopulations, a wide diversity of responses and behaviors can be accurately described without resorting to different mathematical models or adding additional parameters. This seems to be a simpler and more biologically based approach to understanding complexity in seed lots and by extension in diverse biological systems.

Subpopulations are mentioned multiple times in this book where their presence is evident from the data (e.g., Figs. 4.5, 5.6, 6.1, and 6.8). The examples here from seeds illustrate some of the characteristic changes in time courses due to the presence of subpopulations, which can be identified in data across the breadth of biology. This is a critical area for application of the PBT approach, as increasingly individuals are becoming the subject of investigation across levels of organization (e.g., nucleic acids, proteins, enzymes, nuclei, cells, organisms). As a result, topics such as transcriptional or developmental "noise" need to be approached anew with a population-based viewpoint. In multiple systems, hundreds of cell types in what were considered to be uniform tissues are being identified by using single-cell methods (for examples in plants, see Xia et al., 2022; Bawa et al., 2023). These are cellular subpopulations, and their distinct contributions in a developmental pathway, for example, can be identified and quantified using the PBT subpopulation approach.

Many additional aspects of seed biology can similarly be modeled by the PBT approach, as described in more detail in Bradford and Bello (2022). This includes germination responses to temperature, water potential, light, aging (longevity in storage) and hormones as well as physiological responses such as respiration rates (Bello and Bradford, 2016; Bradford and Bello, 2023). As discussed earlier, seed germination is a paradigmatic example of quantal (on/off) behavior of populations of individuals, which is where PBT-based behavior is most obvious. At the same time, the ability of the PBT approach to work well for essentially all aspects of seed biology supports the contention in other parts of this book that the underlying behavior of other complex systems (e.g., cells, tissues) might be better explained by considering them to be populations of individuals organized into distinct sensitivity/response subgroups rather than considering the biological outcome to be examples of undefined molecular or regulatory "noise" around an assumed unique mean value. While stochastic variation may contribute to the generation of sensitivity distributions, biology has adapted by incorporating it into the mechanisms of regulatory systems. Understanding such variation in this way may require viewing it from a population perspective.

Conclusions

The PBT model was originally proposed to describe the germination behavior of seeds in relation to water availability (Gummerson, 1986; Bradford, 1990), and that remains the most fully explored application (Bradford and Bello, 2022). As Eric Roberts (1973, pp. 501–502) precociously noted, "It has [been] suggested that seeds are unpredictable things and consequently it would be a waste of time to try to perceive laws of behavior among such erratic individuals. . . . But other disciplines have shown us quite clearly that, although the behavior of any individual in a population may be quite unpredictable, the behavior of populations of individuals can often be defined very accurately." As predicted, the extended PBT model fits well with biological behavior across all organizational levels of plant life. In fact, much of what I refer to as the "behavior" of microbes or plants is made possible by the responses and adaptations to environmental signals embedded by evolution in sensitivity distributions that do not require consciousness to respond to their environments with agency and intent (Trewavas, 2014; Gilroy and Trewavas, 2023). But what about animals and even humans? Keep your population-colored glasses handy and come along to the next chapter.

6
Animals

Population-based systems in animals

Now that examples of population-based behavior in microorganisms and plants have been described, what about animals? Is there also evidence of population-based mechanisms or behaviors in animals and their cells? Yes, and there is much more evidence than I can consider here, due to the extensive research conducted on animal systems, including humans. I refer the reader to Williams (1956) for a comprehensive overview of the remarkably large variation existing in human anatomy and biochemistry along with precocious insights regarding research and medical practice that align well with the PBT approach. Here, I will present some examples that clearly illustrate PBT-based physiological systems in animals (primarily humans), but I will not attempt an exhaustive catalog. Rather, I challenge readers to put on their population-colored glasses and to consult their favorite scientific journal for instances of data relating to animal molecular and cell biology that could be interpreted in a population-based manner, particularly dose-response curves. They will not be hard to find.

Muscle contraction

An obvious role for populations is evident in the function of muscles based upon the contraction or extension of individual muscle fiber cells. A single muscle fiber cell contains thousands of myofibrils and billions of myofilaments (Frontera and Ochala, 2015). When induced to contract by a nerve signal, the two primary protein components of myofilaments, actin and myosin, utilize adenosine triphosphate (ATP) to power a shortening of the myofilament. Each myofilament can contract by a given amount, generating a force proportional to the number triggered. When you lift a light weight, only the required fraction of the total fibers (or filaments) is called upon. That is, a fraction of the population has a highly sensitive threshold for contraction and will contract first in response to the nerve signal. If the force generated

Populations of Individuals. Kent J. Bradford, Oxford University Press. © Kent J. Bradford (2026).
DOI: 10.1093/9780197846032.003.0006

is insufficient to lift the weight, additional fibers with higher thresholds will be recruited until the strength of their combined contraction is sufficient to the task (or not, if the weight is too heavy). This process will be familiar to anyone who feels what is happening in their muscles when they are lifting a light versus a heavy weight. You can feel your body calling on additional muscles until they are sufficient to lift the heavier weight. If the total population of fibers is insufficient to lift the weight, the only choice is to add more myofilaments to increase the force-generating capacity. Thus, it is apparent why we refer to those who engage in consistently pushing themselves to lift greater and greater weights as "body builders": it is only by adding more actin and myosin proteins in myofilaments to muscle fibers that the lifting capacity (force generated) can be increased (Li and Larsson, 2010), which the muscles will accomplish if they are regularly taxed to contract near their maximum capacity (Lamon et al., 2014). In addition, the motor neurons in muscles show a similar effect, increasing their growth in response to biochemical factors ("myokines") released from the myofilaments and in response to mechanical stretching (Bu et al., 2024). Thus, the muscle system operates on a population basis: the size of the total population of muscle filaments and neurons is regulated in proportion to the demand upon them, as well as incorporating a range of response thresholds among the individual muscle myofilaments and fiber cells. This illustrates the recruitment component of the PBT model, in which additional individuals are recruited to respond as the effector signal increases. The same actomyosin units are also involved in multiple other contractile and movement-associated processes in animals. As *in vivo* and single-cell methods become increasingly available, the ability to control the dosages of effectors and monitor the thresholds of receptors will enable more specific testing of potential PBT behavior in these systems (Nalbant et al., 2023).

Muscle cells also exhibit another feature of the PBT model, which is that the speed of response also increases as the signal level increases. As summarized by Martin (2023, p. 335), researchers investigating nerve impulses from stretch-receptor fibers in frog muscles "found that when the muscle was stretched, the rate of impulses increased in proportion to the degree of stretch. The ensuing rate-coding hypothesis has become a central tenet in understanding the messages carried by individual neurons. For example, high spike rates in sensory neurons engender intense sensations, and if spikes are widely spaced in time, the sensations are correspondingly feeble. Likewise, high spike rates in neurons that innervate muscles generate strong contractions and low spike rates generate only weak contractions." I will discuss nerve cells further later in this chapter, but the signaling frequency from the muscle neurons is more rapid as the effector input (stretching) increases, consistent with the inverse relationship between the amount by which an effector exceeds its receptor threshold and the time to response in the PBT model.

Insulin secretion to control blood glucose levels

Mammals have a well-known system for hormone–receptor interactions controlling the glucose levels in the blood that involves specific cells in the pancreas. These so-called islets of Langerhans are composed of small clusters of individual cells within the pancreatic tissue. Certain ones of these cells (termed β-cells) have the capacity to secrete insulin in response to increases in blood sugar levels, which acts to bring the blood sugar levels back down to the desired level—unless, of course, you have diabetes and the insulin produced is either insufficient or unable to achieve this function. Rodbard (1973) described this fundamental endocrine response system as a "quantal response unit"; that is, individual receptor sites sensing their effector are either occupied or they are not. Based on chemical kinetics, the fraction of receptor sites occupied by the signal molecule (e.g., glucose) will increase as the signal concentration increases, and once a threshold fraction of sites is occupied, the response is elicited. Thus, an individual cell will respond (i.e., turn on insulin secretion) when the number of sites filled exceeds a specific threshold. However, the same response could be triggered if the signal concentration remained lower but the number of receptor sites per cell increased, thereby increasing the sensitivity of the cell to the signal. This simple model illustrates how the sensitivity threshold for cellular response can be adjusted by increasing or decreasing the number of receptors, thereby recruiting different numbers of cells (each exhibiting an on/off or quantal response) in response to the effector level. Similarly, the number of secreting cells will decrease automatically as the insulin has its effect and the glucose level declines. Rodbard (1973, p. 342) modeled this population-based threshold system rather straightforwardly and predicted that "[t]his simple model can be generalized, and may provide a new, fundamental approach to the analysis of endocrine control mechanisms."

Rodbard (p. 359) raised a fundamental question about how such systems respond at the physiological level: "As the response increases, is the fraction of cells responding increasing or is the response of individual cells increasing?". If blood glucose levels increase, do all β-cells respond in an analog or graded manner, each secreting a tiny additional amount of insulin in response to each increment of increasing glucose concentration? Or do the individual cells respond quantally (i.e., in an on/off manner), with their sensitivities varying such that small increases in blood glucose only trigger a small fraction (the most sensitive cells) to respond fully, but the sum of their maximal individual responses also restores blood sugar to its proper level (cf. Fig. 4.4)? What if the person quickly consumed a box of candy and their blood sugar content increased dramatically? In the former case, all the cells would increase their insulin secretion in proportion to the signal level, while in the latter case, the higher glucose level would recruit more cells (having higher glucose thresholds) to each respond maximally, thus ameliorating the hyperglycemia. Which of these mechanisms occurs?

It has long been known that pancreatic β-cells are, in fact, highly heterogeneous (Pipeleers, 1987, 1992; Van Schravendijk et al., 1992) and that the individual islets respond independently and nonsynchronously, "which suggests that each islet is a functionally independent unit" (Valdeolmillos et al., 1996, p. 9). This current understanding was well summarized by Benninger and Hodson (2018, p. 537):

> A β-cell is a terminally differentiated cell that produces and secretes insulin in a glucose-regulated manner. Importantly, β-cells have the ability to adapt to changes in metabolic demand through increased insulin secretion and/or number. In most vertebrate species, β-cells form clusters with other hormone-secreting cells (glucagon-secreting α-cells, somatostatin-secreting δ-cells) within islets of Langerhans. Very early studies of the β-cell assumed them to be homogenous based on a lack of morphological differences. However, detailed studies subsequently determined that there exists a broad heterogeneity in the function of β-cells. These early studies of β-cell heterogeneity are summarized by the landmark review of Pipeleers (1992), which describes with remarkable foresight the presence, characteristics, and role of functional β-cell subpopulations. This includes how dissociated β-cells show functional heterogeneity, with populations of cells displaying higher levels of glucose metabolism, redox state, insulin synthesis, membrane potential, and insulin secretion; that morphological markers (nuclear size, insulin granularity) can differentiate β-cell subpopulations with differing glucose sensitivity and insulin secretion levels; that β-cells show heterogeneous expression of key proteins such as glucokinase (GCK), connexins, or insulin, including spatial variations across the islet; that β-cells with low glucose stimulated insulin secretion preferentially increase in number under development or metabolic stress; and that β-cells vary in their sensitivity to cytotoxic agents.

It is known that "the islet number rather than islet size is the major determinant of total β-cell mass in humans" (Kou et al., 2014, p. 1733). With the advent of single-cell analysis methods, it has become possible to sequence the transcriptomes and analyze the proteomes of individual β-cells themselves in response to increased glucose concentrations (Avrahami et al., 2017, p. 150): "many islet cells . . . display divergent expression patterns within each canonical endocrine cell type. Importantly, we found that insulin transcript levels can vary by more than 100-fold in human β-cells." Furthermore, some of the β-cells can serve as "hubs" that signal other β-cells to also secrete insulin, and there are additional types of cells in the islets as well, including α- and δ-cells (Benninger and Kravets, 2022). At least some of the subtype specification appears to be controlled epigenetically (Dror et al., 2023). Thus, multiple layers of control and of subpopulations of individual cells are involved in regulating blood glucose concentrations, therefore leading to multiple steps where it could fail, resulting in diabetes. A current focus of concern is how such heterogeneity (or lack of it) will impact the use of cultured (i.e., clonal) β-cells in the treatment of diabetes (Pipeleers et al., 2017; Aldous et al., 2023; Tanday et al., 2024).

Many questions remain about how these cell types interact within a given islet and how the multiple islets in the pancreas collectively interact. I have not found any cases where a population-based threshold approach has been used to analyze such data (e.g., Rodbard, 1973), despite the clear initial description of β-cell subpopulations by Pipeleers (1987, p. 281): "Functional analysis of purified islet cells indicated that adult pancreatic B [β] cells are sociologically organised at three different levels. A first level is expressed in the coexistence of functionally diverse subpopulations, a second level is associated to the aggregated state of the cells and a third level is achieved by the synergistic interactions of the B [β] cells with their environmental signals." Pipeleers (1987, p. 282) also noted, based on earlier work (Van De Winkel and Pipeleers, 1983), that "A stepwise increase in extracellular glucose levels resulted in a stepwise increase in the number of B [β] cells with an elevated NAD(P)H and a reduced FAD [flavin-adenine dinucleotide] fluorescence intensity."

To illustrate how the PBT approach can be applied to study these cellular heterogeneities, and how such a PBT viewpoint can contribute to understanding how this complex system may function, I will use an example from Farack et al. (2019) in which they measure the transcriptional heterogeneity of β-cells in individual islets of the intact mouse pancreas. Data from this paper show the number of β-cells in a given islet expressing a specific number of Insulin2 mRNA molecules coding for the insulin protein (Fig. 6.1). The bar graph shows the data as reported in the paper. The authors note that the distribution of the cell numbers is skewed to the right, that is, that a relatively small fraction of cells contains more than twice the median number of Insulin2 mRNA molecules, which they term the "extreme cells" (gray bars in Fig. 6.1). They then compare the cells in the "non-extreme fraction" (black bars in Fig. 6.1) with those in the extreme fraction across a range of other measures. It is not noted or justified why the value of twice the median value (or multiple of the median, MoM) is selected as the dividing line, but the paper focuses almost exclusively on distinguishing those "extreme" cells from the "non-extreme" cells.

Now is one of those times when I will ask you to put on your population-colored glasses and look again at the original data. The shape of the bar graph in Figure 6.1 closely matches the one shown in Figure 3.2, which is simply a log-normal distribution plotted on a linear scale. That distribution matches the data in general, particularly the right-skewed component (green curve in Fig. 6.1). Another alternative is that there were actually two overlapping subpopulations underlying this data, rather than simply an "extreme" tail in a single distribution. I have added two subpopulations in Figure 6.1, a black normal distribution containing 75% of the cells centered near the overall median of 0.85 MoM (with σ of 0.33) and a second gray normal distribution containing 25% of the cells centered at 1.57 MoM (with σ of 0.66). The red line shows the sum of the black and gray subpopulations, which matches with the actual data even better than the logarithmic curve and also explains the right-skewed overall distribution. Of course, this is just me drawing curves and empirically fitting the actual data, but the implications are significant for understanding the behaviors of these cells and the way to pursue studies that can uncover the mechanisms operating in them. If the two-subpopulation model is correct, then instead of there being only 7% of the cells in a distinct higher-producing population, the number may actually be a quarter of the total population (the subpopulation indicated by the gray curve in Fig. 6.1). And in comparing the "non-extreme" and "extreme" populations in these categories, most cells in the second population (i.e., 18% out of 25%) are actually included in measurements attributed to the non-extreme population. In looking at the other parameters measured in these cells, Farack et al. (2019) separate and compare the extreme and non-extreme cells, and statistical analyses indicate significance probabilities (p values) as low as 10^{-15} between the means of the two cell types. But what does this really indicate when the two categories are selected purely on the basis of how far the values are from the median, and statistics only compare mean differences? Of course, the mean of the extreme values will differ significantly from that of the non-extreme ones because they were grouped specifically on that basis. Admittedly, these are early days for understanding how to deal with the masses of data coming from spatial single-cell transcriptomics, but this seems to be circular statistics at best.

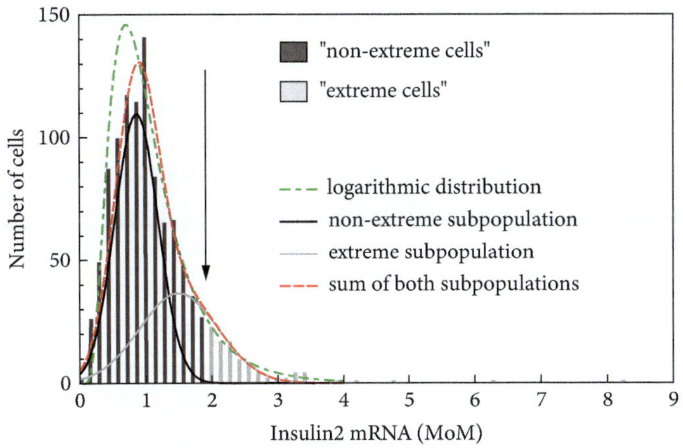

Figure 6.1 Distributions of β-cells in mice pancreatic islets based on Insulin2 mRNA abundance measured by fluorescence probes in sectioned tissues. The bar graph shows a right-skewed distribution of frequencies of different expression levels. The vertical arrow indicates two times the median value (multiple of the median, or MoM) of the relative insulin-encoding mRNA molecules per cell across all cells residing in a single islet. This is taken as a threshold to classify cells exceeding this level as "extreme" cells (gray bars), with only 7% of the total population exhibiting this characteristic compared to 93% of the lower-expressing "non-extreme" cells (black bars). Two alternative interpretations of this data based on PBT principles are shown by superimposing population distributions on this data. The green distribution assumes a single population, but with a logarithmic distribution of relative mRNA production among the cells (see Fig. 3.2); the parameters are $\mu = \log[0.70]$, $\sigma = \log[0.58]$ (plotted on linear scale). An alternative interpretation is that two distinct normally distributed subpopulations are present, indicated by the black and gray curves, with their sum indicated by the red curve. Best fit was obtained when 75% of the cells were in the black population and 25% were in the gray population, with their means and standard deviations (black non-extreme distribution, $\mu = 0.86$, $\sigma = 0.33$; gray extreme distribution, $\mu = 1.57$, $\sigma = 0.60$) empirically fit to maximize the match between their sum (red curve) and the actual data. In this case, the "extreme" subpopulation is part of a larger subpopulation of higher-expressing cells which also overlaps with the lower-expressing subpopulation, and the skewness of the total population is due to the overlapping of the two normally distributed subpopulations.

Adapted with permission from Farack, L., et al. (2019). Transcriptional heterogeneity of beta cells in the intact pancreas. *Developmental Cell* 48(1): 115–125. https://doi.org/10.1016/j.devcel.2018.11.001.

On the other hand, the PBT analysis of subpopulations fully supports the primary assertion that two major categories of β-cells are present. And, almost fortuitously, the population of cells in the selected "extreme" category does contain only cells from the second, highly expressing subpopulation (Fig. 6.1). However, it also suggests that there is considerable overlap and that so-called non-extreme cells might contain a considerable fraction of the extreme subpopulation. The extreme cells in this case are just the higher expressing end of a subpopulation with a greater response or sensitivity to regulatory signals. We could predict, for example, that conducting the experiment at a higher glucose level might shift more cells from the black to the gray population or would shift the gray population to a higher median value; these alternatives could be tested in a dose-response experiment. To their credit, the authors do additional measurements, including spatial localization, but they continue to use the same definition of what the extreme cells are, and so they struggle to document relatively minor differences in some cases. In contrast, recent work by Moussu et al. (2023) in a plant system showed similarly skewed response distributions that could be readily explained by two underlying subpopulations, similar to the distributions shown in Figure 6.1 (see also Fig. 5.12). The resolution that is now achievable using single-cell transcriptomes is likely sufficient to separate such subpopulations (Ma et al., 2024), and therefore will allow exploration of their different responses across concentrations of a given signal factor.

In addition to having different subpopulations with different sensitivity thresholds, this pancreatic system also has multiple distinct cell types and communications among them, including signaling molecules, receptor proteins, and other characteristics (Aldous et al., 2023). This provides the basis for the observed plasticity and heterogeneity, but a PBT approach could bring some order to the situation. A given cell could participate in one subpopulation with respect to glucose sensitivity, and in addition it could belong to a non-identical subpopulation with respect to another regulatory signal. The complexity diminishes if cells exist in structured threshold-based populations with respect to each of these parameters. This would allow subsets of cell populations to be responsive to different physiological inputs, which could in turn affect the mean and distribution of the glucose response of the total population. An approach like the one illustrated in Figure 6.1, along with dose-response experiments with the key regulatory factors, could likely help sort out these complex interactions among individual cells and the subpopulations to which they belong.

Cell division

Another example that clearly exhibits population-based behavior is the regulation of cell division, which is essential for cell proliferation and subsequent development (Fabrèges et al., 2024). Mammary epithelial cells require epidermal growth factor (EGF) to stimulate an extracellular-regulated mitogen-activated protein kinase (ERK) in order to proceed into cell division. This signaling pathway controls cell proliferation, survival, migration, and differentiation. The ERK kinase that is regulated phosphorylates more than 100 substrates, which then determine subsequent cellular phenotypes. To study this process, Albeck et al. (2013) recorded cultured cells over four days in different EGF concentrations and measured the accumulation of ERK protein (pERK) and phosphorylated Rb protein (pRb) (Fig. 6.2A). Both proteins accumulated in a population-based manner; that is, at intermediate EGF concentrations, some cells had responded, while others had not. Summarized across the entire range of concentrations, the responses clearly show that there is a minimum threshold concentration for any response and that increasing fractions of the total population will respond as the EGF concentration increases over a range of nearly three orders of magnitude (Fig. 6.2B). The fact that these results represent behaviors observed over multiple cell cycles in the presence of fixed EGF concentrations indicates that the cells vary in their sensitivity thresholds to EGF, some responding at less than 10 pg/ml, while others require 1000 pg/ml to initiate a response. Thus, these cells respond in a very similar manner as seeds do in response to a germination-promoting hormone (see Fig. 5.8).

The sigmoid increase in responding cell fractions demonstrates one key component of the PBT model: that the population of responding cells exhibits a normal distribution of sensitivity to logarithmic increases in signal concentration. The other component of the PBT model predicts that the rates of response increase as the inducer concentration increases above the individual cells' thresholds. As noted by Sparta et al. (2015, p. 24784), "A challenge in measuring ERK kinetics lies in the use of population level measurements, such as immunoblotting, where data points represent the average of 10^4 to 10^8 cells. Such measurements are degenerate with respect to the states of individual cells because multiple scenarios could give rise to the same population average (e.g., a change in each cell's intensity versus a change in the frequency of 'on' versus 'off' cells) (Purvis and Lehav, 2013). The mean of a population is often a poor descriptor of any individual cell, particularly when cells exist in two or more distinct groups with different properties." Thus, Sparta et al. (2015) used live cell reporters to track ERK pathway activity in individual cells. The full story is quite complex, as the cells can exhibit repeated pulses of ERK activity (and at different frequencies) in response to EGF (Ram et al., 2023a, b). However, their initial responses within each pulse can be followed using fluorescence techniques to monitor ERK activity immediately after application of EGF. The response kinetics are illustrated in Figure 6.3A, showing that the response time is reduced and the maximum activity is increased as the EGF concentration increases (see also Gillies et al., 2017).

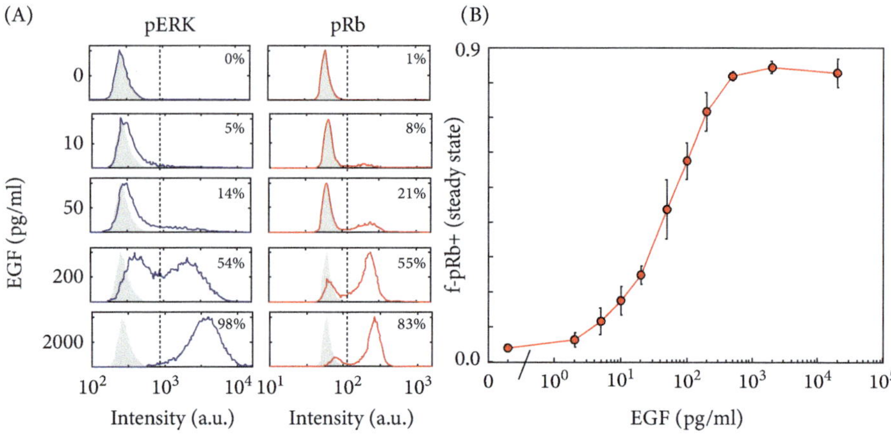

Figure 6.2 Responses of cultured human mammary cells to increasing concentrations of EGF as indicated by increases in ERK protein (pERK) or phosphorylated Rb protein (pRb). Populations of individual cells (200–1000) maintained at constant EGF levels were monitored over 4 days. **A.** Changes in population distributions of the pERK and pRb amounts in the cells. **B.** Summary of steady-state pRb fluorescence at different EGF concentrations.

Reproduced with permission from Albeck, J.G., et al. (2013). Frequency-modulated pulses of ERK activity transmit quantitative proliferation signals. *Molecular Cell* 49(2): 249–261. https://doi.org/10.1016/j.molcel.2012.11.002.

Plotting the percentages of cells responding and their times to 50% of their maximum response gave the patterns illustrated by the data points in Figure 6.3B. These data were fit well by the PBT model (lines in Fig. 6.3B), which indicates that the kinetic aspect as well as the recruitment component of the model both apply to this system.

The population basis of this regulatory system was further confirmed by recent work demonstrating that the cell population is subdivided into multiple subpopulations based upon their own cellular states (Kramer et al., 2022, p. 642):

> In adaptive information processing the response of a signaling node in an individual cell is adapted to the physicochemical state of the cell and its surroundings (here collectively referred to as the cellular state), through mechanisms by which the cellular state controls the activities of signaling nodes. . . . This implies that signaling responses not only capture information about the amount of growth factor a cell is exposed to but also—and perhaps primarily—obtain information about its cellular state. If the activation of different signaling nodes in a network is dependent on different properties of the cellular state then each node would carry partially nonredundant information. As a whole, the network could then generate a multimodal percept that captures a comprehensive picture of a cell's multicellular context and internal state, facilitating accurate and contextual decision-making.

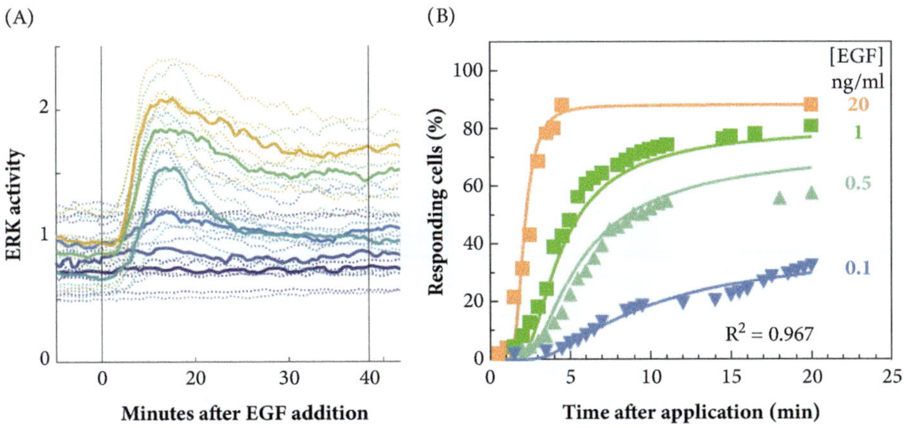

Figure 6.3 Initial responses of ERK activity in cultured mammary cells after application of EGF. **A.** Time courses of ERK activities of individual cells (solid curves = means, dotted curves = ranges) after addition of EGF at the indicated concentrations. **B.** Time courses of the percentages of responding cells at different concentrations of EGF. Cells were defined as responders when intensity increased at least 30% after EGF addition. Cumulative time courses are based on the time to 50% of the maximum response for each responding cell. Solid lines are the time courses predicted by the PBT model with θ_{EGF} = 4.9 log [EGF] min, $EGF_b(50)$ = −0.98 log [EGF], and σ_{EGF} = 0.64 log [EGF].

Unpublished data provided by Breanne Sparta and John G. Albeck, University of California, Davis, presented with permission. PBT time course curves in panel **B** fit by the author.

In agreement with the results shown in Figure 6.2, the responding cell populations shift from low to high levels of pERK, with increasing fractions of the total population responding as the concentration increases (Fig. 6.4A). However, Kramer et al. (2022) further showed that the responses of the individual cells were related to their cellular states (Fig. 6.4B) and that correcting for this clarifies the responsiveness to EGF (Fig. 6.4C). That is (p. 642), "the cellular state has a specific multivariate effect on signaling responses in individual cells." In addition, their model predicts that the responses are most consistent with "switch-like bimodal properties (being either a low or high responder)" being associated with the cell's state (i.e., the subpopulation it is in). These subpopulations can then be represented in terms of their sensitivity differences in response to EGF (Fig. 6.4D), consistent with the PBT concept. Kramer et al. (2022, p. 648) conclude: "Thus, both the position in cellular state space [the threshold sensitivity subfraction] as well as the position in multimodal signaling space [the effector concentration] allows for an accurate prediction of the response of single cells to reenter the cell cycle. . . . We have shown that the heterogeneity in acute signaling responses of individual cells [population-based sensitivity variation] contains partially nonredundant

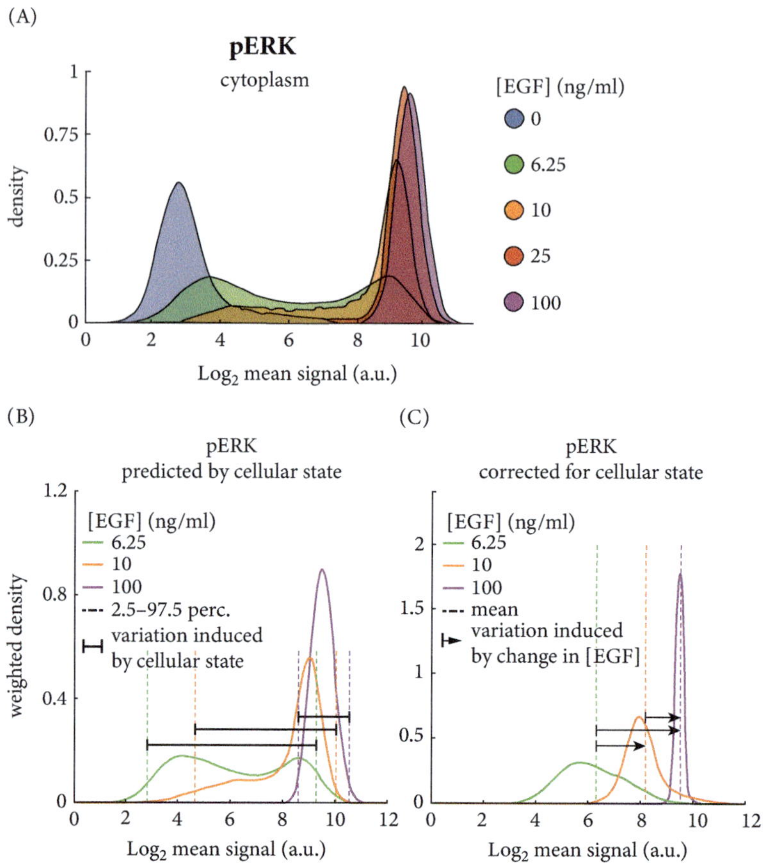

Figure 6.4 Cellular states influence the sensitivity of cell division signaling systems in mammalian mammary cells. **A.** Population densities of pERK in the cytoplasm of individual cells induced with different concentrations of EGF. **B.** Population densities of different subfractions of the cell population indicating the variation induced by the cellular state (i.e., the subfractions with differing sensitivity thresholds in the cell population) in response to increasing EGF concentration. **C.** Population densities induced only by EGF after correcting for the changes in cellular states (sensitivity thresholds). **D.** Cumulative fractions of cells responding to different EGF concentrations, indicating the logarithmic differences in sensitivities among the different cell subpopulations.

Reproduced with permission from Kramer, B.A., et al. (2022), Multimodal perception links cellular state to decision-making in single cells. *Science* 377(6606): 642–648. https://doi.org/10.1126/science.abf4062.

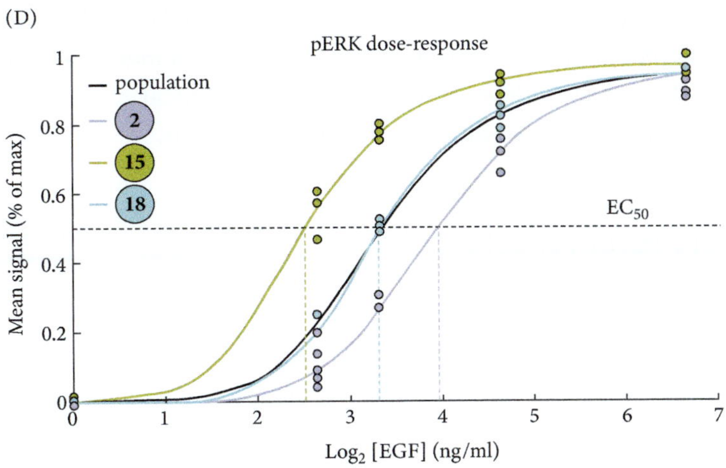

Figure 6.4 Continued

information about the cellular state [subpopulation] that influences the growth factor response." (The bracketed additions to these statements indicate how the PBT terminology would apply to this description.) Despite the obvious population variation demonstrated (Fig. 6.4A), the comparisons and associated discussion largely focus on mean comparisons rather than on the distributions themselves (Fig. 6.4B, C, D). Nonetheless, this study clearly illustrates the power that comes with being able to shift sensitivity distributions in response to biochemical or environmental signals while largely retaining the simplicity of on/off bimodal responses for the individual players, particularly in relation to understanding developmental processes where the primary outcome is differentiation into distinct subpopulations of cells.

Understanding signaling pathways is evolving to take into consideration the spatiotemporal coordination of pulse amplitudes and frequencies in cell populations (DeCuzzi et al., 2024; Huang et al., 2024). Further development of visual biosensors applicable to single-cell technology will undoubtedly strongly contribute to this emerging field (Nakamura et al., 2021). Of particular interest will be the quantitative interactions of cellular waves of spatiotemporal signaling interactions among individuals in interacting cell populations, akin to ripples on the surface of a pond derived from multiple stones thrown into it (Vilches and Dickinson, 2024).

Sensory cells

Interesting examples of cellular development are associated with sensory cells that enable smelling chemicals or detecting light in the environment. In both of these cases, cell populations have been demonstrated to underlie the ability to detect signals from the environment and convert them into information in the brain. Two cases follow where PBT concepts appear to apply and could enhance understanding of the underlying mechanisms.

Olfactory receptor cells

Olfactory receptors (ORs) enable an organism to detect and recognize specific chemicals in their environment. These receptor proteins bind to the unique structural properties of a specific chemical and then cause a signal to be transmitted through a neuron connecting the cell to the olfactory bulb of the brain. In mice there are approximately 1400 functional genes encoding different ORs. In contrast, humans have only a few hundred, indicating our reduced reliance on olfactory signals compared to mice. What is uniquely interesting about this system is the way that the ORs are deployed in the nasal tissue (Fig. 6.5). During development, each olfactory sensor neuron (OSN) cell expresses only one allele of a single OR gene. Only cells expressing that specific OR are then connected together into neurons that lead directly back to the brain. The individual cells therefore are specialized to bind to only a single chemical and to send a distinct signal to the brain when the binding of such a chemical to the receptor protein exceeds its own threshold. That is, the individual OR cells act in a quantal manner, indicating whether a specific chemical is detected at or above its sensitivity threshold level. It is then up to the olfactory bulb to collect the information from all of the OSN cells at a given time and translate it into the identities of the chemicals present in the nose (Monahan and Lomvardas, 2015). The ability of a dog to sniff a handkerchief and then identify its owner by smell is due to its millions of individual sensory cells potentially detecting a thousand or more individual chemicals, with each cell that is expressing the same receptor being connected to the same neuron, which the dog's brain is then able to assemble into the unique olfactory profile of a specific person.

Thus, this system is inherently population based, as it is the population of individual ORs in OSNs that constitute the olfactory sensing system. It is unlikely that a single cell expressing many ORs could interpret such a complex signal internally or transmit it as an electrical (on/off) nerve signal. Evolution solved this problem by distributing the ORs in individual cells that have only one detection responsibility and one unique signal to send. No doubt the signal frequency sent to the brain reflects the number of OSN cells detecting their specific chemical as well as its concentration. How this stochastic segregation of expression into individual ORs in each cell is

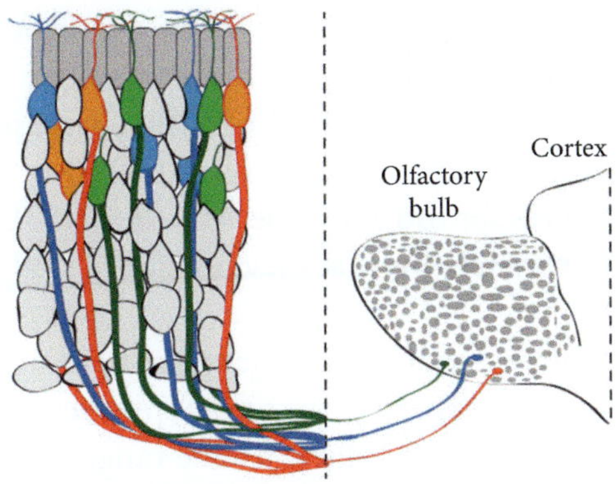

Figure 6.5 Illustration of the structure of olfactory tissues. On the left, the olfactory epithelium shows three distinct types of olfactory sensory neuron (OSN) cells (blue, green, orange). Each of these cells is defined by expressing only one allele of a single olfactory receptor (OR) gene (out of more than 1400 *OR* genes in mice, for example). Cells that express the identical receptor gene (illustrated as being of the same color) connect to the same dendrites and axons and directly to the olfactory bulb in the brain (blue, green, red connecting lines).

Reproduced with permission from Monahan, K., and Lomvardas, S. (2015). Monoallelic expression of olfactory receptors. *Annual Review of Cell and Developmental Biology* 31: 721–740. https://doi.org/10.1146/annurev-cellbio-100814-125308.

achieved and subsequently maintained is itself a complex developmental process (Pourmorady et al., 2024) that enables the sensory system to deliver a unique identification to the brain based on the collective action of all of the individual OR cells. In addition, further thresholds throughout this system may contribute to the overall interpretation. For example, mosquito odorant receptors generate a normal distribution of electrical current increases in individual cells in response to logarithmic increases in odorant concentration (Zhao et al., 2024).

Light-sensing Cells

Light-sensing cells (photoreceptors) in the retina of the eye work together in a similar manner. The most abundant of these are the rod cells, which number over 100 million in the human eye. These cells detect low levels of light, but they do not contribute to color vision, which is handled by the cone cells, of which there are about 6 million in the human retina (Fig. 6.6A). The cone cells detect color by developing different subpopulations that sense light of blue, green and red wavelengths. These wavelength ranges overlap, however, enabling the cells' outputs to combine into signals indicating the full spectrum of colors that we can see when the light is brighter. Thus, as for olfactory sensing, the underlying mechanism for light perception is population based by having specialized cells that represent only a specific component of the total function (Liu et al., 2023). It is also evident in the sigmoid response curves for rod and cone cells (Fig. 6.6B) that their sensitivity thresholds to the wavelengths that they can detect likely vary individually in normal distributions.

In addition, a small fraction of retinal ganglion neuron cells (less than 1% of the photoreceptor cells) expresses the melanopsin receptor and are present in multiple subpopulations that exhibit a wide range of response thresholds covering the entire range from dark to midday (Fig. 6.6). This sensitivity range enables these melanopsin cells to respond to irradiance across orders of magnitude changes in light intensity and to encode circadian rhythms and other internal and environmental signals into animal behavior, called "population radiance encoding" (Dacey et al., 2005; Do, 2019; Liu et al., 2023). The speed of repeated firings of these cells also increases with intensity, providing

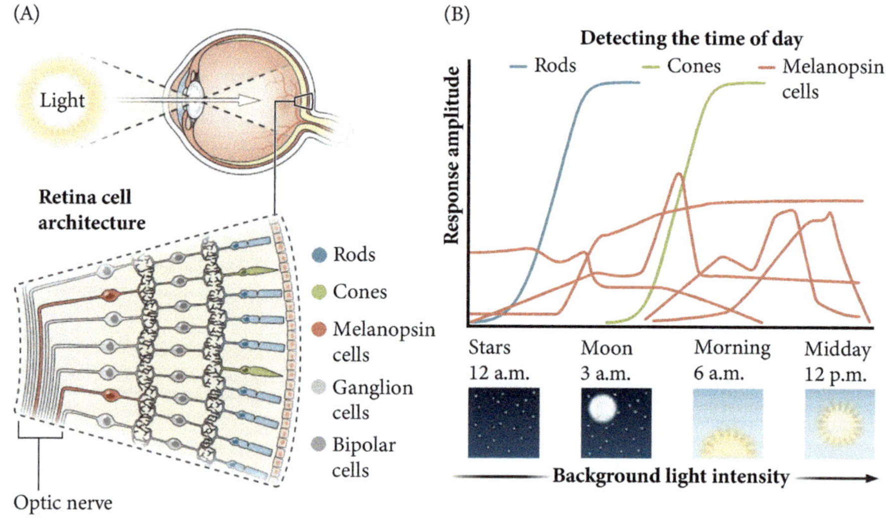

Figure 6.6 Coding of environmental radiance in the retina. **A.** Structure of the retinal lining showing distinct cell types that can detect light and transmit a signal to the optic nerve system. **B.** Rod and cone cells exhibit sigmoid increases in response across specific intensities and wavelengths. In contrast, retinal ganglion neuron cells expressing melanopsin (called melanopsin cells), which constitute less than 1% of the photoreceptor cells, are present in multiple subpopulations that exhibit a wide range of response thresholds covering the entire range from dark to midday. This enables the melanopsin cells to respond to irradiance across orders of magnitude of light intensity, which serves to encode circadian rhythms into animal behavior.

Reproduced with permission from Martin, P.R. (2023). Neurons share an intense load. *Science* 379(6630): 335–336. https://doi.org/10.1126/science.adf9350.

the kinetic response to higher factor levels associated with PBT systems. In addition, subpopulations of these cells that detect the lowest light intensities turn off at higher intensities (Fig. 6.6), conserving energy and clarifying the signal. While the main features of this subdivided photoreceptor system have been known for decades, only recently has the ability to observe individual cells made the subpopulation nature of their functioning evident. This system would be ripe for analysis by the PBT approach, with emphasis on identifying the threshold distributions (for intensity as well as wavelength) of the diverse subpopulations of cells that are present. In particular, it could be interesting to investigate how multiple overlapping population sensitivity distributions come to be represented in the optical nerve and its receptors in the brain. This investigation could not be attempted when only mean values from pooled tissues were available. Now, it is possible to measure the behavior of individual cells in response to gradients of inputs and potentially to follow their outputs into the nervous system. For example, modeling of the detection of motion by 64 individual cells of the fruit fly visual system known to assemble into a "connectome" showed "that such large-scale mechanistic models could accurately make detailed predictions of the neural responses of individual neurons to dynamic visual stimuli, revealing the mechanisms by which computations are performed" (Lappalainen et al., 2024, p. 1139). Threshold-based mathematical models incorporating the diverse cell types interacting in this connectome underlie this computational model. Similarly, recent research on face recognition in primates found that neurons able to identify object boundaries (segmentation) are located in specialized cell modules organized in spatially discrete patches in the visual cortex and that these modules work collectively in cell populations numbering in the hundreds (Hesse and Tsao, 2023). Characterizing this information in terms of population sensitivity threshold distributions, recruitment potential, and kinetic behavior using the PBT concept would be a productive way to proceed.

Developmental timing

The nematode *Caenorhabditis elegans* is a widely studied model system for animal development. Interestingly, a molecular mechanism functioning in *C. elegans* to regulate time-phased developmental expression patterns has recently been described that shares some features with mammalian circadian clocks. As the authors note, "One-fifth of the *C. elegans* larval transcriptome exhibits reproducible, periodic expression patterns that are phase-locked with features of the molting cycle. . . . Oscillating genes are diverse (including [transcription factors], cuticular collagens, proteases, signaling proteins, and miRNAs) with distinct phases of expression throughout each larval cycle. Although this phenomenon was discovered in population-based RNA sequencing studies, understanding the origins and regulation of these patterns requires single-animal or single-cell techniques that can capture continuous expression dynamics within these contexts" (Kinney et al., 2023, p. 2574). The mechanism involves transcriptional pulses of an miRNA regulated by two nuclear hormone regulators (NHRs) that result in developmental stage-specific gene regulatory networks (GRNs). This illustrates how entire GRNs can be controlled by environmental regulation of specific hormonal and genetic (miRNA) components, themselves based on population principles.

Many aspects of this system can be viewed from the PBT viewpoint. However, I will highlight only a single figure as it is related to the prior discussion about the potential heterogeneity of protein molecules themselves (see Fig. 3.5 and the associated discussion). Kinney et al. (2023) demonstrated high-affinity binding between NHR-23 and NHR-85 proteins, which are involved in the temporal phasing of gene expression during *C. elegans* development. The binding curve spans more than three orders of magnitude (Fig. 6.7), which suggests that there is wide variation among the protein molecules themselves in their propensity to bind together. This binding curve was characterized only by a dissociation constant at the median concentration and by an apparent error term (2.2 nM), which is quite small compared to the logarithmic scale on which the data are presented. This is another case where it should not be measurement error per se that we are interested in, but rather in the fact that the binding of these two proteins occurs over a 1000-fold range of concentration between initial binding of some proteins and all proteins being completely bound (saturation). Thus, while these authors have developed remarkable techniques for studying single-cell behavior with respect to transcriptional patterns, and fully appreciate the variation that is evident there, a similar analysis needs to be extended also to the protein level, which is still at the "population-based" (bulk) level noted above for prior RNA sequencing studies.

Nervous system cells

The *C. elegans* nervous system consists of exactly 302 neurons, and yet the organisms can conduct diverse motor programs, including locomotion, feeding, defecation, egg-laying, sleep, foraging and avoidance. Atanas et al. (2023, p. 4134) therefore proposed that in this organism, "it may be feasible to decipher how behavior is encoded across an entire nervous system and how this can flexibly change across behavioral states." These researchers were able to

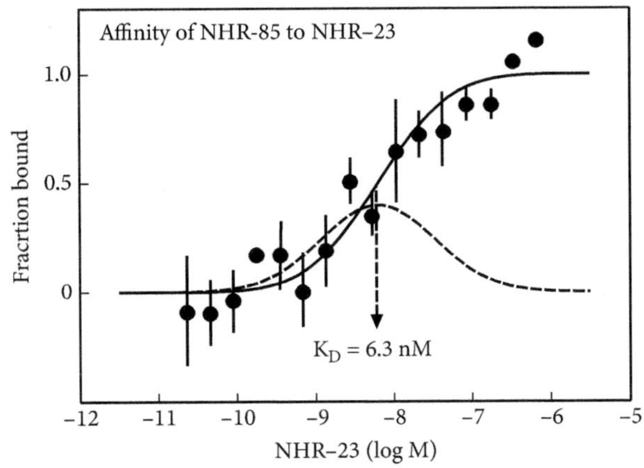

Figure 6.7 Binding affinity of NHR-85 to NHR-23, which controls the temporal expression of a miRNA associated with transcriptional pulses during *C. elegans* development. A log-normal cumulative distribution is superimposed on the graph (black line; in log values, mean = –8.2 log [M], or 6.3 nM, slightly different from the original reported as 5.8 nM; σ = 0.73 log [M]) and illustrates that the binding affinity varies by over 1000-fold between the first proteins to bind and the last ones. This would seem to indicate that variation also exists among the protein molecules themselves.

Reproduced with permission from Kinney, B., et al. (2023). A circadian-like gene network programs the timing and dosage of heterochronic miRNA transcription during *C. elegans* development. *Developmental Cell* 58(22): 2563–2579. https://doi.org/10.1016/j.devcel.2023.08.006.

record brain-wave activity and the motor programs of freely moving animals, and they then developed a probability-based model to connect the neuronal identity and activity to the observed behavior. Among their conclusions were that (1) synaptically connected neurons had more similar behavior than those that were not connected; (2) behavioral information is encoded in "local communities" of neurons across different regions of the connectome system; and (3) such encoding behavior is flexible and is influenced by the behavioral state of the animal. Further (p. 4149), "Depending on the context, the brain may be able to combine different representations to construct new coordinated behaviors." It is clear from this description that even with only 302 neurons, those cells are clustering into subpopulations associated with distinct neural activities; that connections among these subpopulations are critical to their function; and that lived experience can cause changes in the properties of such cells (p. 4149): "30% of the worm's nervous system can flexibly remap" into other subpopulations. In connection with these findings, I point the reader to the remarkable book *Vehicles* by Valentino Braitenberg (1984) in which he cleverly demonstrates how variation in the response thresholds of simple connected signaling systems, combined with the arrangements of their interconnections, is sufficient to describe many of the fundamental features of neural networks. The picture that is emerging from this field suggests that focusing on these response thresholds to different inputs, combined with the population nature of their organization, would be rewarding.

In contrast to *C. elegans*, the 80 billion cells in the human brain are present in at least 100 anatomically distinct regions in which cells have been clustered into at least 31 superclusters, 461 clusters and 3313 subclusters on the basis of single-cell transcriptomes (Lui et al., 2011; Hodge et al., 2019; Siletti et al., 2023). In addition, each neuron can have 1000 synaptic connections, resulting in 80 trillion opportunities for interaction among these cells. Furthermore, "Since neurons interact combinatorially through the synapses they establish with one another, and further so as they interact in networks, the increase in cognitive abilities afforded by increasing the number of neurons in the brain can be expected to increase exponentially with [the] absolute number of neurons, and might even be subject to a thresholding effect once critical points of information processing are reached" (Herculano-Houzel, 2009, p. 9). Neuronal signaling is based on the activation of ion channels, altering membrane ion transport and triggering electrical firing. Some of these receptors are activated by the neurotransmitter glutamate while others that inhibit neuronal excitation respond to gamma-aminobutyric acid (GABA). Recent studies identified plasticity in some of these GABA-responsive receptor proteins with respect to their sensitivities to glutamate and GABA (Piot et al., 2023). These "GABAergic interneurons are famously heterogeneous, and such interactions might be expected to determine the cellular specificity of the plasticity" (Coombs and Farrant, 2023, p. 1364). The known heterogeneity and the characteristic changes in dose-response curves suggest that the PBT approach might be applied advantageously to characterize such plasticity. In addition, Eisen et al. (2024) found that an anesthetic (propofol) that inhibits GABA receptors destabilizes neuronal firing patterns and results in less frequent up-state/down-state bursting patterns rather than the more frequent and coordinated signals present in the conscious state. In response to the anesthetic (p. 2805), "the system's responses to perturbation were much slower than the response during the awake state." This would also be consistent with the slower kinetics predicted by the PBT model as an inhibitory signal exceeds the sensitivity thresholds of the cell population.

A specific example of potential subpopulation analysis is illustrated by Dudok et al. (2024). They described the suppression of the GABAergic responses of specific nerve cells by endocannabinoid (eCB) signaling *in vivo* in mice as they were navigating a linear track. When mammals navigate their environment, specific individual cells (in "place fields") discharge. These discharges were measured *in vivo* as the mice ran on a linear treadmill track. A figure from Dudok et al. (2024) shows the probability density (frequency) diagram (histogram bars) associated with the firing of these specific cells (Fig. 6.8A,B). Curves are drawn through the maxima of these bars (black

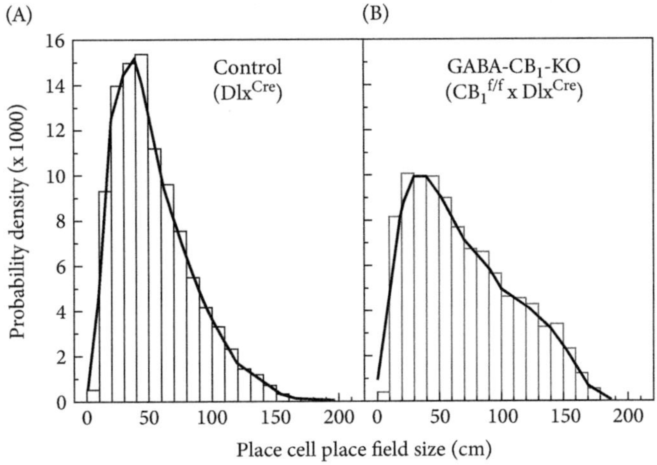

Figure 6.8 Histograms of place field sizes of individual place cells in control mice and after cell-type-specific CB_1-KO [cannabinoid receptor knock-out] in GABAergic neurons (GABA-CB_1-KO). The patterns of histogram bars (gray and red) are shown by the black curves (**A, B**). Subpopulations were constructed using the PBT model whose sum matched these patterns (**C, D**). For the Control subpopulation (**C**), the parameters are: μ =30 cm, σ = 14 cm, and scaling coefficient = 63% for the green subpopulation; μ = 51 cm, σ = 34 cm, and scaling coefficient = 37% for the blue subpopulation; and μ = 54 cm, σ = 30 cm, and scaling coefficient = 26% for the purple subpopulation. The orange dashed curve is the sum of these three distributions (exactly overlapping the original black curve). For the KO mice (**D**), the parameters are μ = 30 cm, σ = 25 cm, and scaling coefficient = 63% for the green subpopulation; μ = 90 cm, σ = 39 cm, and scaling coefficient = 37% for the blue subpopulation. The purple subpopulation is not present (presumably due to its absence in the KO mutant). The sum of the green and blue subpopulations is the orange curve (overlapping the original black curve). Thus, for the KO green subpopulation, the mean remained the same, but the standard deviation almost doubled, while for the KO blue curve, the mean almost doubled, while the standard deviation increased much less. The curves for the sum of the three subpopulations in the Control (**C**) and two remaining subpopulations in the KO (**D**) (orange dashed lines) closely match the original patterns from **A** and **B** (black curves).

Figures in panels **A** and **B** reproduced with permission from Dudok, B., et al. (2024). Retrograde endocannabinoid signaling at inhibitory synapses in vivo. *Science* 383(6686): 967–970. https://doi.org/10.1126/science.adk3863.

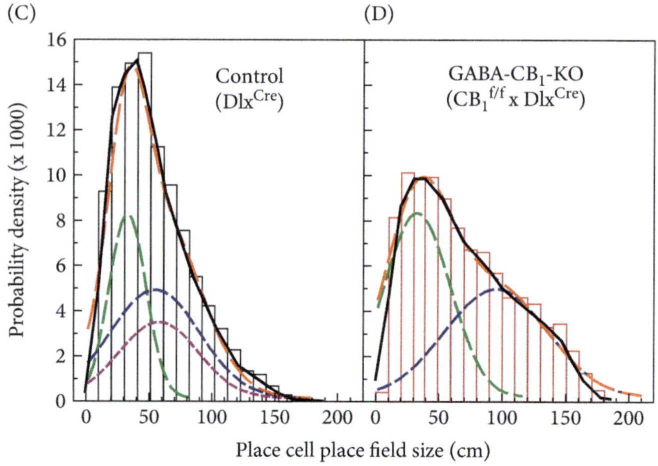

Figure 6.8 Continued

lines) to show a shift in performance after mutating the GABA-CB_1 receptor. However, although the authors show these overall curves, the data are not discussed in terms of possible underlying subpopulations. When viewed through a PBT lens, these histograms seem to clearly show (at least) two subpopulations within the data (compare with Fig. 6.1). PBT-based normal distributions were fit to the data by empirically matching the histogram values to the sum of multiple underlying subpopulations. Two subpopulations could fit the data well (not shown), but scaling coefficients needed to be 63% for each, resulting in a total area under the curve of 126%. This result suggests that a third population was also present representing 26% of the total histogram (purple shortest dashed curve in Fig. 6C). One subpopulation shifted relatively little (green medium dashed line) following knock-out of the *GABA-CB_1* gene, but its standard deviation almost doubled. In the knock-out (KO) mice, another subpopulation (blue short-dashed line) shifted upward in mean (almost doubled), but the variance increased relatively little compared to the control (wild-type) mice (Fig. 6.8C,D). Interestingly, in the KO case, the scaling coefficients for the green and blue curves were 63 and 37%, respectively (the same as for the control condition), and the purple curve was not needed, as this matched the total area under the histogram curve. An obvious interpretation would be that the purple curve represents the cells in which GABA-CB_1 is the primary receptor and that when it is missing due to the knock-out mutation, the other two subpopulations shift their sensitivities in response to its loss.

This is a testable hypothesis that could be explored in further experiments by attempting to identify these subpopulations *in vivo* or by other single cell-based methods. For example, Yang et al. (2024, p. 1478) found "that successive maze traversals [by mice] were tracked by continuously drifting populations of neurons, providing neuronal signatures of both places visited and events encountered," and that learning and memory result from the repeated stimulation of these different neuronal subpopulations. Furthermore, the GABA receptors themselves can be assembled in diverse ways from different subunits (Sente, 2024, p. 739): "[A] cell that coexpressed a set of subunits could simultaneously assemble multiple structurally and functionally distinct receptor subtypes. Further, in this cell, the ratio between coexisting subtypes could be altered purely by changing relative subunit expression levels. Based on our findings and previously known assembly rules, we created a computational model to predict the upper bound on the number of possible $GABA_AR$ structures: 324,727, which is a considerable increase over the previously predicted 800 receptor subtypes (Barnard et al., 1998)!" The key purpose here is to illustrate that population-based thinking can extract additional information and potential explanations from data based on the presence of multiple subpopulations of individuals within the dataset.

This interpretation is supported by other examples of neuronal subpopulations underlying sensory or cognitive attributes. For example, Lavaud et al. (2024, p. 201) described spinal cord cell-type subpopulations that are involved in adaptive learning behavior:

> From adaptation to recall of learned behavior, the neural substrate shifted from a dorsal to a ventral inhibitory PD [progenitor domains] class. We speculate that the spatial position of these neurons, particularly their exposure to somatosensory input, might be a critical factor for this functional division of learning and retention of learned behavior. Dorsal neurons encounter a continuous stream of somatosensory information that can vary from moment to moment, and they may be better equipped to meet environmental demands for adaptive behavior. Conversely, ventral neurons may be more suited to execute appropriate output once familiar cues enter the spinal cord. Transcriptomic findings revealed that dorsal neurons express more plasticity-related genes, whereas ventral neurons exhibit enrichment in genes associated with structural stability (Russ et al., 2021). This further supports the link between the roles of spinal neurons in sensorimotor learning and their spatial positioning. In this study, we identified two distinct neuronal populations as the substrates for limb-withdrawal adaptation.

Similarly, Norman-Haignere et al. (2022) demonstrated that among diverse auditory sounds, specific neurons are specialized for detecting song. On the other hand, neurons can synchronize their activity across multiple subpopulations—for example, to "cleanse" the brain during sleep. Jiang-Xie et al. (2024) described how, during sleep, neurons synchronize their activity into coordinated waves that enable the removal of waste materials from the brain tissue. This demonstrates that neurons can work both individually and as coordinated populations, depending upon the situation.

Neurons associated with memory and learning also exhibit cellular subpopulations with plasticity in their response thresholds (Zhou et al., 2009). Subfractions of such cells with increased excitability (lower thresholds for response) are preferentially recruited into a memory trace (Yiu et al., 2014), exhibiting two key features of the PBT model: sensitivity threshold variation and recruitment phenomena based on such variation. In addition, these properties can be influenced by epigenetic plasticity (Santoni et al., 2024). Liquid–liquid phase separation (LLPS) of proteins into condensates is also involved in regulating synaptic activity and would be another way to compartmentalize subpopulations according to different signals (Guzikowski and Kavalali, 2024, p. 9): "The nano-scale regulation of numerous, distinct signaling pathways supports the functional diversity of a single synapse, vital in all aspects of human life. LLPS is emerging as a key chemical paradigm to explain synaptic protein interactions due to its dynamic nature." As neuronal monitoring methods become more targeted and sensitive, and molecular investigations increasingly move to spatial single-cell methods, it is becoming evident that comparing only means across multiple subpopulations will not uncover the complexity of underlying neural mechanisms that support sensory and cognitive functionality (Wright et al., 2025). Instead, approaching such studies from a PBT viewpoint would contribute to experimental design as well as data interpretation.

Intercellular and interorgan communication

The examples above are only a few of the cells, tissues and systems communicating via diverse signals in a functioning organism. As surveyed by Leslie (2024, pp. 839–840):

> Over the past couple of decades researchers have found that long-distance signaling goes beyond a few glands and organs—the pancreas, thyroid, testes, and ovaries—that spill out familiar hormones such as insulin and testosterone. Most

> organs and tissues now seem to bombard each other with a huge variety of molecular messages. Fat dispatches more than 100 different molecules that may be signals, for example, and muscles send out more than 600. Even bones, long thought to be the strong, silent type, turn out to be chatty, reaching out to the muscles, brain, and other organs. The expanding signaling network also includes the nervous system. Besides identifying a plethora of new molecular signals, researchers have revamped centuries-old maps of the nervous system to include newfound circuitry with unexpected connections. . . . Scientists have just started to catalog the body's cacophony.

It is not unusual that terms such as "cacophony," which is defined as "an incongruous or chaotic mixture," are used to describe such biological situations involving large numbers of potentially interacting signals. However, from a PBT viewpoint, the very existence of such large numbers of unique signaling molecules implies a similar diversity of receptor molecules, each with their own threshold sensitivities to the messages and with variation in these thresholds among individual receptor populations. As has been noted repeatedly in this book, such threshold distributions of signal-receptor populations can bring some order to the apparent chaotic nature of these interactions, while at the same time conveying information about the strength of the signal (and therefore of the appropriate response) and the speed with which such response should be taken. This does not reduce the overall complexity revealed by these recent studies, but it does provide an approach (i.e., characterizing dose-response thresholds for specific messages in cells containing appropriate receptors for them) that can represent large underlying populations of signal/response pairs that automatically interpret the complexity of the signals in terms meaningful for the specific cell types in which the receptors reside.

The tools and data for such analyses are already available. A single-cell-based study analyzed over 2.8 million nuclei from 388 human brains and identified at least 28 neuronal and non-neuronal cell types based upon their transcriptomes (Emani et al., 2024, pp. 3–5). Numerous findings from this study are relevant to the PBT concept:

> We partitioned the variation in expression of each gene according to the relative contribution of individual and cell-type variability while correcting for covariates. This allowed us to determine relative contributions to variability that were based on the function of each gene. For example, brain-specific genes, such as those associated with central nervous system (CNS) morphogenesis and neurotransmitter reuptake, demonstrate a high degree of cell type variability and a lower

> interindividual variation. . . . Conversely, genes associated with common molecular or cellular processes [for example, carbohydrate homeostasis and adenosine triphosphate (ATP) generation] tend to have lower cell-type variation and higher individual variation. . . . Furthermore, an increase in variability is concurrently observed with an increase in cell-type specificity. These patterns held when we jointly considered a gene and its linked upstream regulatory region: A more variably expressed gene is associated with a more variable upstream chromatin region, and both of these are less conserved at the sequence level.

This finding confirms that cell-type variation is important for function, and in fact, it was found to be associated with specific neuropsychiatric disorders (p. 11): "We found 64, 51, 108, and 34 pairs of genes and cell types for schizophrenia, bipolar disorder, ASD [autism spectrum disorder], and AD [and Alzheimer's disease], respectively." Numerous additional examples can be found in a compilation of studies describing cell-by-cell analyses of brain processes (Nusinovich, 2024).

Synapse segregation and biological condensates

Exciting new discoveries are also being made concerning how variation might be maintained even within a single cell, such as a single dendrite of a nerve cell, in which excitatory and inhibitory synapses do not overlap, but instead are spatially segregated. These regions of postsynaptic density (PSD) are distinct for excitatory (ePSDs) or inhibitory (iPSDs) synapses. This segregation is essential for brain function, but it is unclear how it is accomplished at this scale. It has now been shown that this separation occurs because "ePSD and iPSD components segregate into distinct condensates through phase separation" (Zhu et al., 2024, p. 920). Phase separation, in which cellular components (e.g., proteins) in the cytoplasm form condensates (essentially membrane-less organelles) distinct from the liquid cytoplasm, has been identified as a major factor in cellular function, such as for ribosomes (Banani et al., 2017; Ramat and Simonelig, 2022) and stress responses (Romero-Perez et al., 2023; Solis-Miranda et al., 2023). In fact, the nucleolus in the nucleus is such a condensate (King et al., 2024). Protein structure, often associated with intrinsically disordered proteins (IDPs) or regions of such sequences within functional proteins, enhances the ability to form such liquid–liquid condensates (Emenecker et al., 2021). Zhu et al. (2024) showed that the proteins associated with ePSD and iPSD synapses formed distinct condensates, keeping them separate within the nerve cell. Segregation of protein types in condensates is also involved in synaptic junction formation: "phase separation of pre- and postsynaptic scaffolds exhibits a high degree of specificity that could account for the amazing selectivity of the protein compositions of synaptic junctions" (Zhang, X. et al., 2025, p. 328).

This feature of condensates provides an excellent opportunity for operation of the PBT scheme in cellular function, particularly the recruitment aspect (see also the Plant Development section in Chapter 5). The ability of specific proteins to form condensates, presumably restricting their physiological function, yet being available to resolubilize into the cytoplasm and resume activity, provides a mechanism for "storing" fractions of a protein or enzyme population when physiological demands requiring it are low, without accumulating excess amounts in the cytoplasm (see Fig. 8.2; Klosin et al., 2020). For example, processes that operate on a circadian clock, such as sleep in animals or photosynthesis in plants, may use condensates as a mechanism to sequester biomolecules during part of the diurnal cycle (Field et al., 2023; Zhu et al., 2023). Condensates can also segregate ions in the cytoplasm, affecting electrochemical equilibria, pH, and cellular signaling mechanisms (Dai et al., 2024). As the ability to form condensates can be varied by the amino acid sequence of IDP regions incorporated into functional proteins, this could be another mechanism by which a population of transcription factors or enzymes might be sequestered and then returned to activity in response to changes in time of day, signaling factors or metabolite levels. It was recently demonstrated that despite their 3D structural variation, it is possible to design binding proteins that can interact specifically with the intrinsically disordered regions of proteins, enabling more detailed studies of their conformations and functions (Wu et al., 2025).

Immune cells

As cells associated with immune responses must deal with a wide variety of developmental threats and pathogens, it is not surprising that they exhibit high cellular diversity. As reported by Shalek et al. (2013, pp. 236, 237):

> Recent molecular studies have shown that, even when derived from a seemingly homogenous population, individual cells can exhibit substantial differences in gene expression, protein levels and phenotypic output, with important functional consequences. . . . Here we use single-cell RNA sequencing to investigate heterogeneity in the response of mouse bone-marrow-derived dendritic cells (BMDCs) to lipopolysaccharide. We find extensive, and previously unobserved, bimodal

> variation in messenger RNA abundance and splicing patterns, which we validate by RNA-fluorescence *in situ* hybridization for select transcripts. . . . [M]any genes . . . although highly expressed on average, showed significantly greater levels of heterogeneity, with expression levels deviating 1000-fold between individual cells in extreme cases."

This corresponds to the 1000-fold variation in enzyme production by individual seeds discussed previously (cf., Fig. 5.9; Still and Bradford, 1997). Consistent with a PBT approach to interpreting cellular diversity, Shalek et al. (2013, pp. 238, 239), further note:

> Bimodality among highly expressed immune response genes may reflect the presence of distinct cellular subtypes or stochastic differences in the activation of regulatory circuits. . . . These results highlight that single-cell RNA-Seq can sensitively distinguish between closely related, yet distinct developmental states, even within the same cell type. . . . [T]he low population-level expression of [some] transcripts may sometimes reflect high expression in a small subset of cells as opposed to uniform levels of low expression.

Immune cells in the lymph nodes (LNs) also exhibit various types of memory in relation to antigens that have been encountered previously. However, how such memory is embodied and retained in the regulatory T cells (T_{regs}) that comprise the effective immune response has been unclear. It was recently demonstrated that there are distinct subtypes of T_{regs}, called "circulating" and "resident" cells. Kaminski et al. (2023, p. 1) noted that "whereas most T_{regs} in LNs recirculate, 10 to 20% are memory-like resident cells that remain in their respective LNs for weeks to months. Single-cell RNA sequencing revealed that LN-resident cells are a functionally and ontogenetically heterogeneous population." A resident, clonal and heterogeneous subpopulation of cells can persist in the LNs and perform a memory function, while diverse circulating T_{regs} can address current immunity functions in the circulatory system. This work documents multiple examples of distinct population-based sensitivity distributions to different inputs. In addition, immune responses vary widely among individuals and across the lifetimes of individuals, indicating their inherent complexity and sensitivity to the conditions encountered (Metcalf et al., 2025).

Cancer

As would be expected from the work described above regarding immune cells, a large and growing body of work is utilizing single-cell approaches to investigate cellular heterogeneity in both development and disease, particularly cancer. It is not possible here to do a deep dive into this rapidly progressing field, so only a few examples can be given. For example, after describing a simplified approach to obtain single-cell transcriptomes and using it to identify multiple resistance phenotypes within a background of drug-sensitive cells, Clark et al. (2023, p. 1564) noted that "scRNA-seq provides unparalleled insight into cell heterogeneity but remains underutilized in many settings." For example, Kirschenbaum et al. (2023) showed how to identify specific immune cells related to cancer and follow their developmental trajectories over time. This enables the identification of cell subpopulations during development or in response to chemical or other factors. Different signal molecules identified to be associated with specific biochemical processes and cell populations could then be followed spatially and temporally for their *in vitro* and *in vivo* effects. As they concluded (p. 160), "Single-cell technologies opened a new chapter in biology by large-scale phenotyping of cells as the minimal unit of life. Analyzing millions of cells has provided seminal insights into cell states across diverse tissues and pathologies." Nyman et al. (2023) extended this approach by using "spatially aware deep learning" to identify enormous microheterogeneities in the histological features of tumor cell populations that were associated with their sensitivities to different treatment strategies (p. 8): "Simultaneous quantitative measurements of key tumor and microenvironmental properties that represent distinct modes of oncogenesis, evolution, and immune evasion may unlock insights in ccRCC [clear cell renal cell carcinoma] biology and potential modes of patient stratification." This was confirmed by Sinha et al. (2024), who used single-cell transcriptomics of tumor cells to predict their responses to treatments and could classify patients into responder and non-responder groups based on the single-cell transcriptomes of their tumors.

A view of the future of this field can be seen in the work of Ferri-Borgogno et al. (2024), who conducted molecular, metabolic and subcellular mapping of 3-D spaces in ovary and ovarian cancer cells using multiomics analyses. This work not only demonstrated the variation that is present at each of these levels of complexity but also identified new candidate genes based on their ability to better identify distinct cell populations that interact with the immune and cancer cells. As they concluded (p. 20):

> Mapping tumors and precancerous lesions is critical in personalizing medical treatments to the environments found in each patient and holds the best promise for the optimal outcome for that patient. This means an understanding of which cells are present, what their functional statuses are, and how they are energized is critical to understanding how a patient's disease will progress and in identifying the

aforementioned optimal treatment. . . . As we dive deeper into the data mapped to the central cartesian coordinate map we can further correlate these analytes with each other and with cellular functions such as movements and communications, as well as clinical outcomes and responses to therapies. This will function on multiple levels and allow for the development of not only better model systems to test novel therapies, but may also suggest novel therapies, which will improve the survival rates of cancer patients.

Aging

Cellular population dynamics are also associated with aging. A recent study used single-cell methods to characterize sex-related differences in mice across tissues and organs, aging-associated cellular states, and population dynamics of cells during aging, "identifying >300 distinct cell types and >3000 cellular states" and "clustering >200 cell subtypes undergoing marked aging-associated population changes" (Zhang, Z. et al., 2025, p. 10). The authors concluded:

> We identified a complex and dynamic landscape of aging at the cellular level, uncovering >10 main cell types and >200 subtypes undergoing significant age-associated depletion or expansion. Although some cell types, such as lymphocytes, have previously been documented to expand with age, our study has uncovered a range of rare cellular states that remain underexplored. . . . These aging-associated cell types were consistently observed across sexes and even genotypes, underscoring their potential as antiaging targets for further therapeutic exploration.

This assessment illustrates the value and potential outcomes of focusing on cellular diversity and its underlying population structure in future studies of organismal growth, development, senescence and aging.

Conclusions

As we discovered for plants, animals across the evolutionary spectrum also exhibit many examples of PBT behavior. The revolutionary implications of single-cell methods to examine individuality at the cellular, nuclear and molecular levels across spatial and temporal scales fit better with the PBT approach than does dismissing the variation discovered as being "noise." It is certainly noisy in the sense that a lot is going on simultaneously, just as we are constantly exposed to a massive amount of electromagnetic radiation across a spectrum of frequencies. However, that does not mean that we cannot tune a radio, TV receiver or cell phone to a specific source frequency to extract meaningful information from it. The different frequencies are essentially different subpopulations into which the overall spectrum can be partitioned (i.e., Fourier transform analogy) and to which different receivers have sensitivity. While the overall population of radiant energy wavelengths may be noisy and incomprehensible when pooled together, it all makes sense when specific signals and receivers coalesce into efficient sensory and communication networks. Given that a major driver of animal pathobiology research is the discovery of potential applications for human health, appreciation and application of the PBT concept can take advantage of the individuality inherent in biological systems and can avoid inferences based on pooled samples and statistical methods focused only on mean separations rather than on the underlying subpopulation compositions and functions.

7

Developmental Plasticity, Ecology and Evolution

Phenotypic variation and adaptive plasticity

Ecology and evolution are generally discussed in terms of genetics. This approach is reasonable, as genetic variation is the basis of evolution, and ecological success is determined by the genetic capacity to survive and reproduce in a given environment. However, phenotypic variation, or plasticity, particularly in response to environmental stresses, is also involved in adaptive capacity and reproductive success. According to Klingenberg (2019, p. 2):

> Each component of variation has a direct cause or origin that constitutes the input of variation into the developmental system. These direct causes are inducing the differences between different instances of a phenotypic trait, which make up the respective component of variation. For instance, genetic variation originates from individuals that differ by carrying different alleles for relevant genes, phenotypic plasticity results from individuals experiencing different environmental conditions, and developmental instability results from random variation in developmental processes (developmental noise), which produce deviations in each developing structure from the norm expected for that genotype and environment. Such variation is the input for the developmental system, which translates it into the output of the observable phenotypic variation.

Note that Klingenberg (Klingenberg and Nijhout, 1999; Klingenberg, 2003) and others (Polak, 2003) define the stochastic variation component as "developmental noise" or "developmental instability." Often, developmental models incorporate "standard Gaussian white noise" in order to accommodate stochasticity without actually utilizing it as an underlying mechanistic feature (Pezzotta and Briscoe, 2023). This definition of non-genetic variation as noise or instability follows from the emphasis on genetic variation (and its responsiveness to environmental conditions) as the only meaningful (i.e., heritable or controllable) source of biological variation. While genetic,

Populations of Individuals. Kent J. Bradford, Oxford University Press. © Kent J. Bradford (2026).
DOI: 10.1093/9780197846032.003.0007

environmental and stochastic variation contributes to observed phenotypic plasticity, the assumption is that without the stochastic component, all individuals sharing the same genotype in the same environment would develop into an identical "target phenotype" exhibiting little or no variation among individuals in the population. This implies that the stochastic component is both inherent and uncontrollable noise that occasionally breaks through into an observable phenotype. Let's look a bit deeper into this issue.

Ubiquitous stochastic variation

First, it must be recognized that this "target phenotype" without stochastic variation does not exist except in an assumed idealized and simplified world. This is clearly described by Lajus et al. (2003, pp. 345, 360):

> Developmental instability is commonly accepted to be a property of individual organisms. According to this view, developmental instability of an organism reflects its inability to produce an "ideal" form under a particular set of conditions. . . . Similarly, developmental stability refers to the ability of a developing organism to produce a consistent phenotype in a given environment. . . . Developmental instability can manifest itself as a difference between symmetric structures . . . within a single individual. Moreover, it can result in differences between individuals, and can be analyzed when variation caused by environmental diversity and genetic heterogeneity is negligible. For instance, when one compares twins, or clones cultivated under standard conditions, any difference must be caused by developmental instability. . . . One can easily imagine, and even create, a situation where genotypic and environmental variations are nil, but one cannot create a situation where stochastic variation is nil. Stochastic variation is ubiquitous. . . . Any developing organism exhibits instability of development and thus stochastic variation, which cannot be attributed to either genotypic or environmental diversity. . . . Moreover, the nature of stochastic variation as a manifestation of phenotypic variation was more or less ignored by the researchers dealing with it. The ignorance of stochastic variation originates, in part, from non-acceptance by many researchers of fundamental sources of unpredictability in the surrounding world in general, and in biological systems in particular. Those who studied developmental instability considered it merely a characteristic of a population, and they focused on the association between developmental instability and either environmental parameters or genomic coadaptation. In contrast, researchers working with phenotypic variance considered only genotypic and environmental sources of variation. Because genetic variation was the main focus of interest, all nongenetic variation

> was attributed . . . to environmental variation. . . . Instability of development must be considered a third independent source of phenotypic variation, equal to those of genotype and environment.

West-Eberhard (2003, p. viii) made a similar point:

> Already a focus on developmental instability has caused evolutionary and developmental biologists to look hard at the unordered side of variation. . . . I predict that this new focus on the sources of variation will lead to a profound change in how biologists think about the causes of variation and the origins of novelty. Instability can be seen either as innovation or as noise. Labeled as noise, it has been on the junkheap of both developmental and evolutionary biology for decades. This may be necessary in the early stages of understanding, when even the regularities in patterns of genetic variation and evolution are poorly understood. The study of instabilities marks a new maturity, and an exciting frontier. It coincides with the progress currently under way in molecular biology, where genome studies have also brought new interest in the sources of variation and change.

In the two decades since these statements, it is clear that techniques for studying the sources of biological variation and change have become much more sophisticated, documenting the remarkable variation in physiology and development even among individual cells in a single tissue (Shalek et al., 2013; Jean-Baptiste et al., 2019; Schlegel et al., 2024; Schott et al., 2024). As Hallgrimsson et al. (2019, p. 71) have framed it:

> At the heart of this issue is the fact that quantitative genetics and developmental biology are concerned with different kinds of questions and have different definitions of ostensibly the same phenomena. While quantitative genetics is concerned with partitioning phenotypic outcomes into cumulative, statistically-defined phenomena, developmental biology is concerned with physical mechanisms such as specific molecular, cellular or tissue-level interactions. While the genetics of a population of clones is singularly uninteresting, its developmental biology may be fascinating. As a result, these fields produce different kinds of explanations. Quantitative genetics defines most phenomena at the population level and is fundamentally concerned with variation. The mechanisms of interest to developmental biology occur in individuals and, at least under the current paradigm, variation is more often a nuisance than a direct object of study.

The PBT model is based on the conscious and purposeful recognition of the stochastic component of biology, as revealed through the ubiquitous presence of systematic variation among presumably genetically identical individuals (e.g., cells within a single organism). The PBT approach is therefore relevant to the process of changing our understanding of inherent physiological and developmental variation from a noisy nuisance into an integral feature of biological reality and functioning. While the consequences and applications of this developmental viewpoint for evolution and ecology cannot be covered exhaustively here, I will give some examples of how adoption of a popupulation-based viewpoint could occur and why it would be useful.

Modeling developmental variation

Klingenberg (2019) has presented a coherent model to account for the different sources of developmental variation. This model is primarily focused on what is termed "fluctuating asymmetry," or the phenotypic differences that can appear on the left and right sides of an individual organism, such as in the wings of insects on each side of their body. This fluctuating asymmetry (FA) across diverse organisms has been used as a model system for studying developmental instability, as an individual organism has a single genome and its environment is also singular (Gangestad and Thornhill, 1999; Nijhout and Davidowitz, 2003; Zakharov et al., 2020). The developing organism can be viewed as collecting inputs from gene allelic variation, environmental variation and stochastic noise, which feed into the developmental process to determine the actual phenotype, given the genotype and the environment in which development is occurring (Figure 7.1A). Of particular interest here is Klingenberg's developmental model based upon this concept, which incorporates thresholds (Fig. 7.1B). It is founded upon a diffusion-based mechanism in which a developmental signal molecule (e.g., morphogen) diffuses away from a single source, resulting in its decreasing concentration with distance from the source. This decaying signal then interacts with downstream receptor molecules or cells and triggers a response when its level is above the response threshold at their location. Note that the threshold value is considered to be a constant across all receptor molecules or cells. The model also posits that the receptor cells can produce a "background" level of the signal molecule, but this background level also is taken to be constant across all the receptor molecules/cells (Fig. 7.1B). This produces a model that predicts variation in phenotypes that is dependent upon the genetic alleles present (controlling the source signal) and varies across environments (Fig. 7.1C). The slopes (tangents) of these response curves at each point reflect the sensitivity to developmental noise for each genotype of a given regulatory locus.

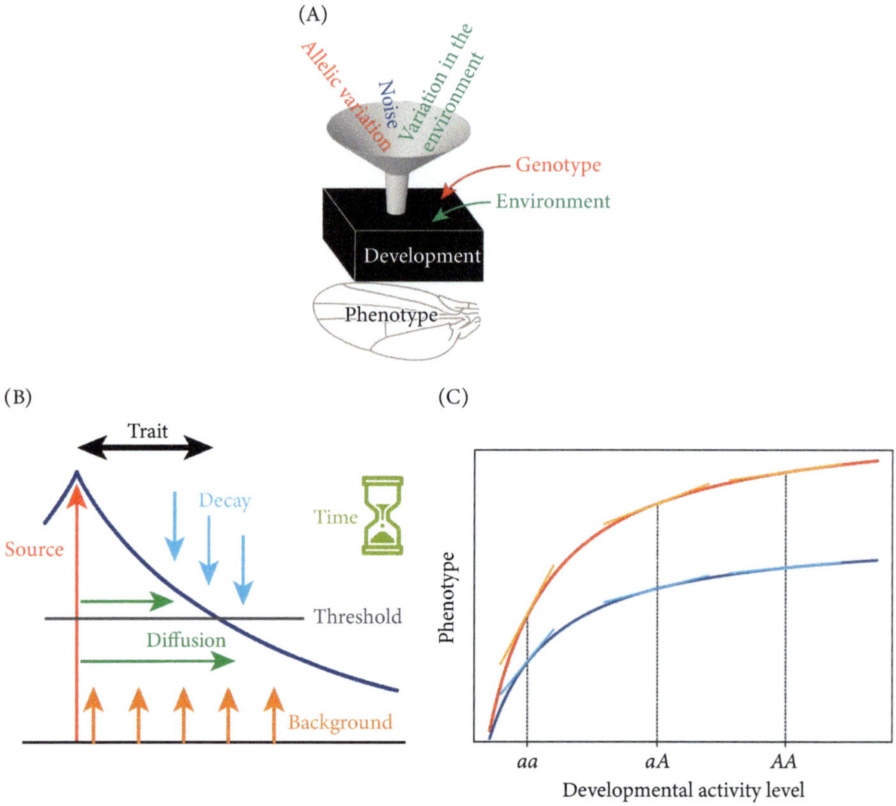

Figure 7.1 Inputs into phenotypic variability. **A.** Genetic, environmental and stochastic variation (noise) contributing to the phenotypic plasticity of development of a specific genotype in a given environment. **B.** Diffusion-based model of changes in signal (e.g., morphogen) levels over time as a function of diffusion away from a source (decay), with receptor molecules or cells responding when the signal level exceeds their common threshold. These cells can also contribute to a background level of the morphogen signal, which also is considered to be constant among the receptor molecules or cells. **C.** Predictions of this model for phenotypic variation in relation to allelic dosages for a specific locus associated with the phenotype in two different environments (red and blue). The tangents shown on the curves indicate the sensitivity to developmental noise for each gene dosage in each environment.

Reproduced with permission (Open Access) from Klingenberg, C.P. (2019). Phenotypic plasticity, developmental instability, and robustness: The concepts and how they are connected. *Frontiers in Ecology and Evolution* 7(56): 1–15. https://doi.org./10.3389/fevo.2019.00056.

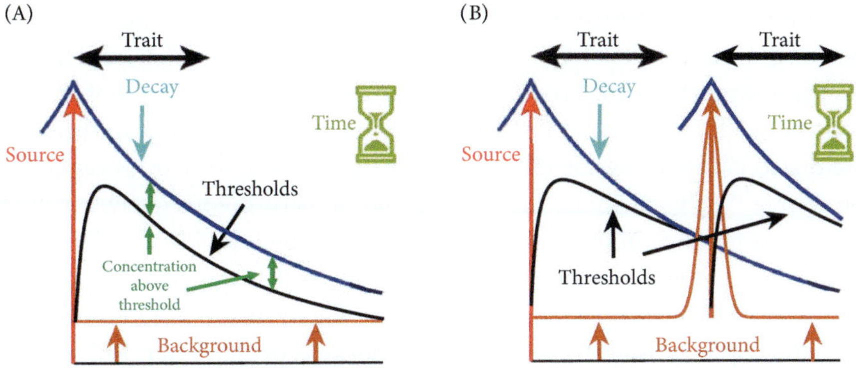

Figure 7.2 Revision of Figure 7.1B to accommodate PBT principles of threshold and background variation in the receptor population. **A.** Instead of a common response threshold, the downstream cells exhibit variation in their thresholds (black curve). The curve reflects the shape of a logarithmic range of sensitivity plotted on a linear scale (see Fig. 3.2). In this case, this results in a stable signal level in excess of the threshold along the diffusion path. Thus, biological variation in individual thresholds could counteract the decline in signal concentration to give a uniform trait signal across a physiological or cellular distance. **B.** Similarly, instead of a fixed ability to produce the signal themselves (termed background), different downstream cells could also vary in their expression of the signal depending upon their location or sensitivity to other signals. For example, when the signal strength declines below the response threshold, that might remove repression and reproduce the original signal source. Thus, the interplay of a physical process (diffusion) with population thresholds could result in consistently spaced repeating developmental units, which are common across biology.

Adapted with permission from Klingenberg, C.P. (2019). Phenotypic plasticity, developmental instability, and robustness: The concepts and how they are connected. *Frontiers in Ecology and Evolution* 7(56): 1–15. https://doi.org./10.3389/fevo.2019.00056

Thus, random variation has a much larger impact on phenotype in the *aa* genotype than in the *AA* genotype in this example.

This model shares some aspects with the PBT model. Both are based upon sensitivity thresholds to developmental signals. Both result in models predicting the developmental progression over time in different environments (compare Fig. 7.1C with Fig. 2.4B). The major difference between the models is how stochastic variation is incorporated into the model. In the developmental instability model, variation is due to the source strength of the signal and subsequent diffusion to receptor molecules or cells that all share the same threshold value for that signal (Fig. 7.1B). This is consistent with current models in enzyme kinetics, which assume that all enzyme molecules are identical and variation in reaction rates with concentration is primarily due to physical limitations such as diffusion (but see Fig. 3.6). In contrast, in the PBT model, the receptor molecules or cells vary in their thresholds (Fig. 7.2), and this represents one mechanism through which the stochastic component of variation is introduced. However, as illustrated in Figure 7.2A, variation in thresholds could also compensate for the decline in signal strength with distance from the source. Such sensitivity variation could, at least theoretically, result in a constant difference in the signal level relative to the individual cell's sensitivity threshold at any distance from the source. In this way, biology can be in control even though physics may determine the diffusion kinetics of the signal (although it is doubtful that diffusion per se is primary in cellular tissues, where distances are relatively large and multiple options exist for controlling both movement and concentration, such as compartmentation, specific directional transporters, etc.).

In addition, there is no reason to assume that all the receptor cells are identical in their own "background" production of the developmental signal. Why is the ubiquity of stochastic variation ignored in this case? Such a distributed signal production capacity would represent a distinct population-based feature that will interact with the distribution of threshold sensitivities and the decay of the signal from its primary source (which itself may be a unique cell population rather than an individual). A specific theoretical example of this interaction is illustrated in Figure 7.2B, which shows one possibility for when the signal level declines below the response thresholds due to diffusion. Rather than the effect disappearing, it could be that loss of that signal triggers the creation of a new source. That is, the presence of the signal represses its own production, keeping the background synthesis low in the vicinity of the source. However, once the signal level falls below a threshold level, that repression would be removed, and those cells could reproduce the initial source of the morphogen. This would be a simple way to space repeated developmental elements, such as lateral roots along a main root axis or serrated leaf margins in plants or sequential development of components such as body segmentation in insects or teeth in developing animal jaws. Spatial specificity and kinetic variation in plant stress signaling waves for reactive oxygen species and salicylic acid have been demonstrated experimentally using *in vivo* nanosensors (Ang et al., 2024). We also can consider the likelihood that distinct feedback signals from the "downstream" individuals could diffuse back to the source regions to modify their signal production (Wu et al., 2009; Mukherji et al., 2011; Nijhout et al., 2014). These feedback distributions also likely include stochastic variation at the most basic biological level, due to the intrinsic "noise" or "instability" in biological processes, but at the same time they potentially convert much of what would be termed developmental instability at the individual level into tissue-level (i.e., population mean) developmental stability (Ebert and Sharp, 2012; Shalek et al., 2013).

It is also important to note the image of the sand clock in Figures 7.1B and 7.2, indicating that development happens over time. However, unlike a clock, developmental time has a physiological basis rather than a fixed physical basis (Bradford and Trewavas, 1994). In addition, as has been illustrated throughout this book, the rates at which biological processes progress are influenced by the magnitudes of the signals regulating the process or developmental transition of interest. The greater the difference between the signal and the receptor threshold, the more rapid the response (Fig. 2.4); this is occurring independently for each individual in the population. Or as Watahiki and Trewavas (2019) expressed it, increasing concentrations of an effector shorten or "compress" time. This is analogous to temperature, particularly for microorganisms, plants and cold-blooded animals, where higher temperatures increase metabolic and developmental rates, shortening the time required for a process to occur, at least up to a point at which elevated temperatures can become damaging (Alvarado and Bradford, 2002). Higher metabolic rates effectively compress more activity (respiration, metabolism, development) into a shorter time period (Walde et al., 2025). This is exactly the kinetic component of the PBT model, and the shape of the threshold sensitivity distribution relative to the signal level determines the shapes of the response curves (e.g., Figs. 2.4, 5.12). This finding led Bradford and Trewavas (1994) to suggest that life actually runs on "biotime," which can speed up or slow down as needed in response to internal or external signals. This is evident in hibernation among insects and animals and dormancy in plants, as well as in circadian cycles in both plants and animals. But it is also present at the molecular level, not only by affecting diffusion rates, but also as described in the PBT model as a function of signal levels and response thresholds that also are sensitive to environmental conditions.

Thus, I do not concur with the interpretation that the presence of stochastic variation per se implies "instability" of biological processes, or at least the negative connotation of that term. In fact, the overlapping sensitivity distributions to promotive and repressive signals as envisioned in the PBT model (Fig. 5.8) would function as stabilizing factors enabling homeostasis at the population level, while providing a mechanism for maintaining individual phenotypic plasticity and adaptive potential. It is evident now, as illustrated in previous chapters, that at the molecular and cellular levels, variation of individuals in defined population distributions is the norm, not the exception. Thus, our assumption of uniform, unchangeable biological building blocks that are subject solely to the binding laws of physics must go the way of Newtonian fundamental ideal particles in the face of quantum uncertainty (McFadden and Al-Khalili, 2014; Yvert, 2014; Frank et al., 2024). Just as Newtonian principles still work at larger scales, focusing on population means rather than distributions has been able to describe many aspects of biological behavior, but as we dig down into deeper and deeper underlying layers of biological function, the uncontrollable stochastic variation associated with thermodynamics plays an increasing role (Parthasarathy, 2022). However, as posed in the initial chapters of this book, are we to assume that the incredible diversity and success of biology has occurred without evolving mechanisms that could deal with such uncontrollable physical stochastic variation, and in fact incorporate it into fundamental homeostatic mechanisms while maintaining adaptive plasticity in the face of constant environmental variation? The complex mechanisms for error-checking and correction of DNA molecules are evidence that biology is capable of evolving systems to control random error such as DNA breakage and copying errors (Waterworth et al., 2019). Nonetheless, we have perhaps placed too much emphasis on the stability of this genetic repository essential for hereditary transmission of a common set of genetic tools encoded in DNA and too little focus on recognition of how the inherent stochastic variability in biological processes is dealt with at the transcriptional, protein, cellular and organismal levels (Mukherjee, 2022; Arias, 2023).

A step toward resolving this issue would involve relatively small (but critical) modifications of Klingenberg's diffusion-based model (Fig. 7.1B), as shown in Figure 7.2. Diffusion and decay of signals both within and among cells and tissues are no doubt involved. However, allowing the thresholds to vary among individuals in the receptor populations would automatically incorporate the stochastic variation that is considered to be "noise" or "instability" in the Klingenberg model. The result is still phenotypic plasticity, but it is an intrinsic component of the developmental mechanism, rather than being what is left over after genetic and environmental effects have been accounted for. A distribution of threshold sensitivities rather than a single constant value would incorporate the stochastic effects occurring at the molecular level directly into the model. Similarly, allowing variation in the background production of the signal would mean that there would not need to be a single distinct source of the signal molecule, but rather a community effort that would allow further influence by the source strength, such as via distinct cell populations or transmission pathways. It is also possible to imagine a parallel system in reverse for producing regulatory feedback molecules that would have their own effects on participants upstream in the diffusion pathway. Such population-based feedback systems are critical components of nonlinear regulatory behavior (Gilroy and Trewavas, 2001; Johnston and Bassel, 2018; Watahiki and Trewavas, 2019; Gilroy and Trewavas, 2023).

As Klingenberg (2019, p. 2) noted, "There is not always a straightforward relation between the inputs of variation into the developmental system and the phenotypic outputs." There are other processes that can modify or suppress phenotypic variation, such as through canalization, buffering and robustness (Waddington, 1942; Debat and David, 2001; Nijhout, 2002; Félix and Barkoulas, 2015; Hallgrimsson et al., 2019). It is worth referring to Waddington's original description of canalization, or the channeling of development into pathways resulting in phenotypic traits that would be favored by selection (Waddington, 1942, p. 563):

> Naturalists cannot fail to be continually and deeply impressed by the adaptation of an organism to its surroundings and of the parts of the organism to each other. These adaptive characters are inherited and some explanation of this must be provided. If we are deprived of the hypothesis of the inheritance of the effects of use and disuse, we seem thrown back on an exclusive reliance on the natural selection of merely chance mutations. It is doubtful, however, whether even the most statistically minded geneticists are entirely satisfied that nothing more is involved than the sorting out of random mutations by the natural selective filter.

Waddington's developmental example was the formation of callus pads on the skin of ostriches that appear even in developing embryos, well before the friction and pressure on specific locations of the skin that usually are associated with formation of such pads. He further stated (p. 564), "If we suppose that the callosities, when they were first evolved, were dependent on the environmental stimulus, then the evolution appears as a readjustment of the reactivity of the skin to such a degree that a just sufficient thickening is produced with the normally occurring stimulus." Such a readjustment in "reactivity" would be identical to a change in the sensitivity thresholds as defined in the PBT model. In fact, Waddington further considered that (p. 564) "the magnitude of the response would not be proportional to that of the stimulus; there would be a threshold of stimulus, above which the optimum (that is, naturally selected) response would be formed. In so far as the response became canalized, the environment would be acting as a switch." Again, this is directly in line with the PBT concept that the response of an individual often is more like a light switch than a dimmer. A given individual responds in a switchlike (on/off) manner when its response threshold is exceeded; increasing signal concentration above the threshold affects the speed of the response, but amplification of the signal is also due to recruitment of additional responders in the population by the higher concentration rather than to specific modulation of the response magnitude in each individual (see Fig. 4.4). This explanation is presented explicitly by Waddington in his concluding paragraph (p. 565; italics are added here for emphasis):

> Summarizing, then, we may say that the occurrence of an adaptive response to an environmental stimulus depends on the selection of a suitable genetically controlled *reactivity* in the organism. If it is an advantage, as it usually seems to be for developmental mechanisms, that *the response should attain an optimum value more or less independently of the intensity of the stimulus received* by a particular animal, then the *reactivity* will become canalized, again under the influence of natural selection. Once the developmental path has been canalized, it is to be expected that many different agents, including a number of mutations available in the germplasm of the species, will be able to *switch* development into it; and the same considerations which render the canalization advantageous will favour the supersession of the environmental stimulus by a genetic one. By such series of steps, then, it is possible that an adaptive response can be fixed without waiting for the occurrence of a mutation which, in the original genetic background, mimics the response well enough to enjoy a selective advantage.

Thus, the roots of the PBT concept were present at the early stages of considering how current environmental conditions that affect survival can be incorporated into genetically transmissible traits. What was missing is an appreciation of the role of individuals (whether molecules or cells) and the variation in their sensitivity thresholds ("reactivity") as integral parts of the fundamental mechanisms of population-based perception, adaptation and intentionality that underlie all levels of biology (Trewavas, 1999; Bradford, 2018; Gilroy and Trewavas, 2023; Vane-Wright, 2023). In addition, this would particularly be the case when different components of the system (signal generation, transmission, reception, reaction) each are themselves parts of distinct signaling and response distributions. There is a widely repeated meme based on the ancient myth that the earth sits on the back of a turtle (e.g., Sapolsky, 2023). The next question is: what does that turtle sit on? And the answer is: another turtle; and so on: "It's turtles all the way down." This is the way that I view the roles of individuals and their cooperative action in hierarchies of related subpopulations: "It's populations all the way down" (see Fig. 2.7). Fully modeling this complex and interacting hierarchy mathematically across the entire range of biology is beyond my capabilities, but it should be rather straightforward for a mathematician (particularly with an

AI assistant), as the basic PBT model is quite simple, with clearly identified parameters for optimizing the fit to observed data. It is also rather simple to conceive that while this model retains individuality, plasticity and adaptability at a given level of activity, it could effectively transmit the integrated response to the next higher level of organization.

Phenotypic plasticity and PBT models

As has been the case throughout this book, the primary interest here is in the role of nongenetic variation in biological functioning. In addition to stochastic variation, phenotypic plasticity is also a recognized source of nongenetic variation among individuals. Franks et al. (2014, p. 124) defined phenotypic plasticity "as the ability of a genotype to express a different phenotype under different environmental conditions." Similarly, Klingenberg (2019, pp. 3, 5) noted that the phenotypic plasticity "of a trait is variation of the target phenotype for a given genotype in different environments. In particular, a reaction norm characterizes plasticity by describing how the target phenotype for a specific genotype varies as a function of an environmental variable. . . . The observed variation in a phenotypic trait is therefore the combined effect of the developmental noise and the sensitivity or robustness of the developmental system." Thus, in addition to genetic and stochastic variation, organisms have the capacity to vary their current phenotypes in response to environmental conditions, based upon their sensitivities to such conditions and the range of responses available to them. It is this underlying sensitivity threshold distribution that is identified by the PBT approach. It assumes that total potential variation is bounded by the genes available for use by an organism, but the fraction of that potential variation (including downstream variation in and modification of proteins) that is actually expressed as a phenotype is dependent upon the sensitivity thresholds of the individuals in the relevant population to changes in environmental factors, where "environmental" has a broad interpretation, being, for example, the accessible diffusion limits at the molecular scale, other cells in a tissue, daily environmental changes for a circadian trait, or seasonal variation for ecotypes or species. In particular, the PBT approach focuses attention on both the mean and the variance of responses associated with the sensitivity of a reaction to a given effector. This is supported by Félix and Barkoulas (2015, p. 494), who noted that "[a]nalysis of the mean and variance along dose-response curves would facilitate interpretation of an observed change in variance."

The relationships of the individual to the population and to phenotypic plasticity are the subject of some disagreement. According to Nijhout and Davidowitz (2003, p. 6), "A target phenotype is the property of an individual. A target phenotype will change when the genes or alleles change (by mutation, recombination, drift, or selection) and when the environment changes." On the other hand, Klingenberg (2019, p. 3) states, "There are important theoretical and experimental situations, however, where multiple individuals share the same genotype and environment and therefore those individuals also share the same target phenotype. With modular organisms, such as plants, it may also be useful to apply the concept of target phenotype to discuss variation among repeated parts within an individual, e.g., variation among leaves or flowers in response to their position on the plant (Herrera, 2009). It therefore seems preferable not to tie the concept to a particular level, such as the individual." As argued previously, the term *target phenotype* has only theoretical meaning, as it assumes a situation with no stochastic variation, whereas stochastic variation is ubiquitous. Thus, as a theoretical concept, multiple individuals could share an identical target phenotype, but in reality, the target phenotype will be defined primarily as the mean phenotypic value of a trait in a population composed of individuals, while attributing the variation among those individuals not exhibiting the mean value to "noise," or stochastic variation. In addition, the definition of "individual" seems to refer only to a complete organism, as leaves or flowers of a given plant are apparently not allowed to be called "individuals" in their own right (but see Fritz et al. (2018)). That has not been the approach here, as populations of individuals have been described at every biological level, from molecules to cells to tissues to organisms to species (Fig. 2.7). Thus, I do not agree that individual phenotypic variation should be defined only at the entire organism level, but rather at each organizational or functional level in biology.

In fact, it is the thesis of this book that population-based properties encompassing the diversity of individuals are recapitulated at every functional level in biology (Fig. 2.7). From the PBT point of view, the level of organization does not matter; what is relevant are the responses that occur in relation to changes in effector levels (e.g., environment in this case) and their timing. If the source of variation is genetic, phenotypic plasticity will often be embodied in multiple genes of small effect. It is possible to identify these genetically, such as by quantitative trait loci (QTL) or genome-wide association (GWAS) analyses, if a specific phenotypically identifiable trait (at any level of biological organization) is directly caused by (or linked genetically to) a given heritable locus. However, the fact of phenotypic plasticity even in the absence of genetic variation (within an individual organism) shows that physiological and developmental variation allows plasticity in response to environmental inputs.

One possible explanation of this effect is the multiplication of gene families encoding proteins of similar or identical function (Groot Crego et al., 2024). Gene families coding for proteins of similar function with distinct but overlapping promoter thresholds sensitive to different physiological or environmental signals are common in biological evolution (Gildor and Smadar, 2018). In many cases, these are differentiated by the timing and magnitude of their expression developmentally (e.g., transcription factors) or in response to environmental signals (e.g., heat or light), contributing to phenotypic plasticity. In addition, the increasing appreciation of the extent of epigenetic control of transcription multiplies the avenues for effecting and modulating specific gene expression patterns among existing genes (Dey et al., 2015; Keung and Khalil, 2016; Candela-Ferre et al., 2024). MicroRNAs (miRNAs) provide another mechanism to regulate gene expression into proteins (see Fig. 5.3) and may function in part to buffer stochastic changes in transcription within and among cells (Mukherji et al., 2011; Siciliano et al., 2013; Sekar et al., 2023). In many cases, it is the population distribution of responses that can be generated by this complex underlying system that is of interest (e.g., Figs. 6.1, 6.8).

This is where PBT approaches can assist the strictly genetic approaches that run out of signal when a given gene's input is at the "noise" level. However, the PBT model does not require a specific mechanism, such as a diffusional signal or a unique receptor. By working specifically at the population level but appreciating and valuing the contributions of individuals in that population, the PBT approach can be applied at any biological level (Donohue et al., 2015; Bradford, 2018). It is a rather extreme extrapolation to use biophysical diffusion rates or membrane potentials, which undoubtedly underlie biological functioning, to explain why you preferred eggs rather than pancakes at breakfast today (Frank et al., 2024). It is similarly unreasonable to assume that your choice was predetermined at the Big Bang in a purely deterministic universe (Sapolsky, 2023). New, purely biological, emergent traits appeared as the underlying biological systems became more sophisticated at perceiving their surroundings and enabled organisms to have intentionality and agency in their activities (Gilroy and Trewavas, 2023; Kauffman and Roli, 2023; Mitchell, 2023). This is a component of plasticity, or directed variation, in response to predictable factors in an organism's environment. If they are successful, particularly if the changed factors are permanently rather than only temporarily required in a new environment, the preferences generated will become ingrained epigenetically or by selection of genetic variation

(Stinchcombe and Kirkpatrick, 2012; Noble et al., 2019; Stamp and Hadfield, 2020).

Some comments by Radersma et al. (2020, pp. 360, 366, 368) are relevant to this issue:

> Phenotypic responses to a novel or extreme environment are initially plastic, only later to be followed by genetic change. . . . These results suggest that plasticity is a persistent source of developmental bias that shapes how plant populations adapt to environmental change, even when plasticity does not constrain how populations respond to selection. . . .
>
> Animal[s] and plants are inherently responsive to their environments. Such "plasticity" can evolve, which explains why, for example, plants that grow in shade develop large leaves; modifying leaf size helps the plants to capture the right amount of light. While it is uncontroversial that responses to the environment can be adaptive, it is debated whether or not plasticity also directs genetic evolution. That is, do individual responses to a novel or extreme environment influence how populations adapt to this environment? Here, we analyze data from 34 studies of plasticity and local adaptation in plants to test if this is the case. Our results strongly suggest that locally adapted traits are modified versions of ancestrally plastic responses, making plasticity appear to "take the lead" in adaptive evolution. The immediate responses to the environment were sometimes more extreme than the phenotypes of locally adapted plants, which can give the false impression that plasticity and evolutionary adaptation oppose each other; in fact, truly maladaptive plasticity was rare. While the signature of plasticity persists during local adaptation, trait combinations could be modified independently of plasticity during genetic evolution. It thus appears that plasticity is an important source of adaptive developmental bias in plant evolution, which may facilitate rather than hinder adaptation. . . .
>
> If those plants encounter conditions that are sunnier, drier, or windier than what they are adapted to, this mechanism may not only enable individuals to persist, but will consistently provide natural selection with a particular subset of the many possible phenotypes that could be adaptive in those conditions. Through this logic, populations that have evolved adaptive plasticity should continue to adapt by modifying the mechanisms and traits that are environmentally responsive, rather than inventing new adaptations. As a result, evolution will tend to go where plasticity leads. The extent to which populations really do adapt to novel or extreme environments using the same mechanisms and traits that they employ in adaptive plasticity is poorly understood, however. . . . We therefore anticipate a growth in the number of studies that combine mechanistic studies of the developmental basis of phenotypic variation with measures of selection in ecologically relevant settings.

Similarly, Capblancq et al. (2020, p. 259) noted that:

> Standing (i.e., pre-existing) genetic variation is not the only factor that determines population persistence, and future efforts should focus on integrating genomic prediction models with other ecological and physiological processes and data sources. For example, individuals might tolerate exposure to rapidly changing environments through physiological plasticity. . . . Measuring plasticity under multiple relevant climatic environments and combining organismal data (morphology, physiology, etc.) with variation at the transcriptomic and epigenomic levels could prove a more nuanced view of vulnerability to climate (mal)adaptation that could improve predictions based on genetic offset alone.

Population-based threshold models represent an approach to the "mechanistic studies of the developmental basis of phenotypic variation" and "measuring plasticity" that is called for here, as they specifically focus on and quantify phenotypic variation and its responses to input factors, whether molecular, hormonal, environmental or genetic. As noted by Tsimring (2014, pp. 25–26): "The new frontier in studies of biological fluctuations is to figure out how the noise that originates at the gene expression level propagates up the developmental ladder to the levels of tissues, whole organisms and populations. Along this ladder there are multiple layers of regulation, both intracellular (transcriptional, post-transcriptional and epigenetic) and extracellular that can shape the characteristics of noise. Analyzing such complex non-linear, non-equilibrium, multi-scale systems not only poses serious technical challenges but calls for new conceptual paradigms." Part of that new conceptual paradigm is the possibility that life managed to "harness stochasticity" in relation to cellular epigenetic mechanisms during development, as discussed by Jablonka and Lamb (2020). It is interesting to note that Noble and Noble (2018) used the same term to describe the ability of emerging organisms at the origin of life to harness stochasticity as part of their fundamental functional mechanisms (see Chapter 3). From a PBT viewpoint, harnessing stochasticity would be accomplished not by suppressing variation per se, but rather by adapting to such variation and building it into regulatory systems through its capacity for recruitment and kinetic (timing) responses to changing signal levels. Jablonka and Lamb (2020, pp. 40–41) also recognize the value of thresholds in epigenetic inheritance and adaptation: "If an induced change in gene expression leaves a mark that alters the threshold of future responses to that inducer, epigenetic inheritance of this primed state is likely to be beneficial: it does not lead to inappropriate responses, because the environmental stimulus is still necessary, but the speed and efficiency of response are heritably altered. The advantage of rendering the expression of a gene more or less likely following earlier induction in an organism's parents or more distant ancestors is similar to the advantages of neural learning, especially sensitization and habituation."

Developmental threshold models

Thresholds and bimodal (quantal) responses have been incorporated directly into developmental models, as in Waddington's description of canalization and in the agent-based modeling methods (Railsback and Grimm, 2019; Zhang and DeAngelis, 2020). Biological development occurs over time, as cells differentiate into distinct cell types and tissues. Developmental time accumulates until it exceeds a threshold, at which time the next developmental event will occur (Donohue et al., 2015). In plants, this developmental event could be the transition from vegetative growth to flowering, or in insects, the change from larva to adult. As biological processes are sensitive to temperature, such developmental time is often expressed in thermal time units (degree-days), or the temperature in excess of a base temperature (minimum or threshold temperature, T_b) multiplied by the time (t), or $(T—T_b)\ t$. In many cases, plotting the time to a developmental transition on a thermal time basis will normalize the effects of different temperatures on the developmental time required (Alvarado and Bradford, 2002). In other cases, a period of time under specific conditions may be required to achieve the developmental threshold, such as for hydrated chilling to remove seed dormancy or time below a certain temperature to induce flowering in plants (vernalization). In these cases, once an individual seed or plant has accumulated enough chilling time, it will transition into germinating or flowering once the temperature increases. In this sense, developmental "thresholds" can represent minimum accumulations of developmental time that allow transition to the next developmental phase, where "threshold" has an extended meaning as a "door" through which the cell or organism passes to go to the next stage (Fig. 7.3; Donohue et al., 2015). In other cases, the signal can cause a more abrupt change in development. For example, outside of equatorial regions, the daily duration of light (photoperiod) is a reliable signal of the season of the year. Many plants have adopted this daily photoperiod as a signal for triggering transition to flowering and reproductive development at the appropriate time (Gendron and Staiger, 2023). Some plants will flower after the photoperiod decreases below a threshold length (i.e., nights are long), while others will flower only when the photoperiod exceeds a minimum threshold (i.e., nights are short), which coordinates the developmental transition to the season appropriate for flowering of the species. In some species, this can be a very abrupt transition, where only a single day of the correct daylength will trigger flowering, while in other cases repeated exposure to the inducing daylength is required. Knowledge of the photoperiod requirements of *Poinsettia*, for example, allows growers to manipulate their growing environment to bring the plants into flower to coincide with the winter holidays when they are sold. In addition, signals can be combined to adapt plant development to specific environments. For example, *Arabidopsis thaliana* grows across a wide range of latitudes, from the Mediterranean to the Nordic regions. Across this wide range of habitats, the combination of environmental inputs (temperature, photoperiod) and sensitivity thresholds (seed dormancy, vernalization requirements) adjust germination and flowering times such that seed maturation and dispersal can occur in either the spring or the fall, but seed development will coincide with a time period when the average temperature is between 14 and 15°C in the location in which the seed is developing (Springthorpe and Penfield, 2015). For further examples of developmental thresholds influencing plant behavior, see Donohue et al. (2015) and Auge et al. (2019).

The key to applying PBT models or developmental threshold approaches to ecological and evolutionary studies is to measure the responses of organisms to different levels of the factor of interest. Such "dose-response" curves measured on individuals in a population are the basis of determining the mean and standard deviation of the responses within the population across factor levels. This is often used for temperature, as it is easy to control in the laboratory and biological processes often show clear changes in rates of developmental progression across temperatures (Burggren and Mendez-Sanchez, 2023). However, many other environmental factors influence development, often simultaneously and interactively, and the systematic testing of these factors to develop sensitivity threshold distributions and kinetic response data over time to fully apply PBT models will be challenging, particularly across multiple genotypes (Xue et al., 2024). These interactions are particularly difficult to quantify in natural environments, where many environmental factors can change simultaneously and stochastically, but it can be approached, for example, by common garden experiments in different environments or by sampling across a consistent environmental gradient, such as altitude (Gremer et al., 2020; Exposito-Alonso, 2023).

In addition, the typical experimental protocol of using different, but consistent, environmental conditions during growth to assess phenotypic plasticity may be misleading or inadequate to predict organismal responses

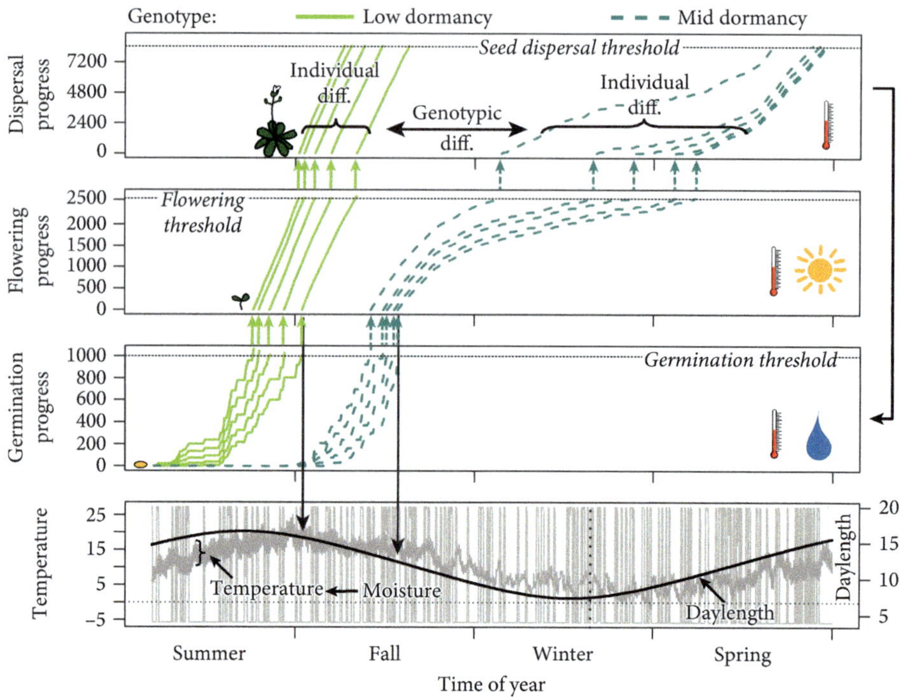

Figure 7.3 An integrated life cycle threshold model illustrating genotype-specific phenologies and generation times across a species range. The general structure of an integrated model that links three developmental threshold models across life stages is shown, based on the annual plant *Arabidopsis thaliana*. Environmental inputs over the course of a year are shown in the bottom panel: photoperiod (smooth black line); precipitation events (vertical gray lines); and temperature (continuously varying gray lines). Beginning with the seed stage (second panel from bottom), progress toward germination proceeds in response to the environmental inputs of water availability and temperature. Green versus blue lines indicate different "genotypes" with different parameter values for dormancy that reflect natural allelic variants known to occur in *A. thaliana*: low dormancy (green) and intermediate dormancy (blue). The multiple lines for each genotype represent random variation among individuals within a genotype in their rates of progress toward germination. Once the developmental threshold has been attained for an individual seed, it germinates and becomes a vegetative rosette (transition to the next higher panel). These rosettes experience different environments because the seeds they came from germinated at different times. Rosettes progress toward the transition to flowering in response to temperature, photoperiod and winter chilling. Once the thresholds for these factors are reached, flowering occurs (transition to top panel), and seeds develop and mature at rates that depend on temperature. Finally, seeds are dispersed (final threshold at top of diagram) and the cycle begins again. Note how the phenotypic differences among individuals within and between genotypes change over time in response to the environmental cycles.

Reproduced with permission from Donohue, K., et al. (2015). Applying developmental threshold models to evolutionary ecology. *Trends in Ecology & Evolution* 30: 66–77. http://doi.org/10.1016/j.tree.2014.11.008.

to stochastic weather extremes exceeding historical boundaries (Burggren, 2019). While mean global temperatures may only increase 2°C due to climate change, the temperature extremes occurring at a given location will be much larger, potentially extending beyond the range of adaptation or survival of some species. Assessing the range of such extreme plasticity would require not only modification of experimental designs (Burggren, 2019; Burggren and Mendez-Sanchez, 2023), but also changes in data collection protocols, with an emphasis on identifying the fraction of individuals adapting or surviving at the extremes of the sensitivity distribution, not just pooled means and statistical tests of whether the means are distinct. Instead, the emphasis should be on quantifying the standard deviations of the populations and the factor values more than two standard deviations from the mean, representing the most sensitive and most resistant 2% of the population (Fig. 2.2). The climatic variables themselves may need to be expressed in frequency terms in order to enable them to be applied in relation to organismal responses to crossing their environmental adaptability thresholds (Dillon et al., 2016). In addition, as organisms can adapt to environmental changes by shifting their thresholds over time (Dahal and Bradford, 1994; Stinchcombe and Kirkpatrick, 2012), experiments may need to assess such plasticity (or "flexibility") both before and after a period of such stress and/or adjustment (Burggren and Mendez-Sanchez, 2023). While increasing the complexity of such experimentation, it will be critical to assessing the threats of climate change to the biosphere, and the PBT approach of quantifying the full distribution by assessing individuals, not just the pooled population means, will be an essential tool for implementation.

Seed dormancy and bet hedging

As discussed in Chapter 5, seeds are the paradigmatic example of PBT behavior (Bradford and Bello, 2022, 2023). They are individual organisms; they are dispersed individually; their propensity to germinate or remain dormant is affected by genetics and by the environment; and they monitor and respond to multiple input factors before transitioning from dormancy to germination (Finch-Savage and Footitt, 2017; Reed et al., 2022). These effects are especially evident in wild annual species, where survival is dependent upon proper timing of flowering, seed dispersal and germination in relation to the environment (Vidigal et al., 2016; Hawkins et al., 2017). However, within this context, a remarkably wide range of dormancy and germination strategies can be successful (Adondakis and Venable, 2004; Baskin and Baskin,

2014; Baskin and Baskin, 2021). In addition, particularly in variable and often unfavorable environments, seeds provide the mechanism for ecological "bet hedging," or favoring some survival over all environments at the cost of less than maximal reproduction in optimal conditions (Venable and Lawlor, 1980). Seed dormancy is a key mechanism for bet hedging in plants, as it may reduce the fraction of seeds that would potentially germinate under better conditions, but it is more likely to have some seeds persist during the worst conditions (Gianella et al., 2021). Seeds of native species are often exquisitely attuned to their environments, moving into and out of dormancy while in a seed bank that constitutes the future of the species (Donohue et al., 2010; Rubio de Casas et al., 2015; Finch-Savage and Footitt, 2017; ten Brink et al., 2020). Successful reintroduction of species requires not only seeds from genotypes adapted to the environment of the location, but also timing of seed distribution to match seeds' expectations or requirements (Baskin and Baskin, 2020). In many cases, very low success rates are observed in such restoration efforts, while conversely, seed dormancy characteristics may play a critical role in the competitive advantages of invasive species (Meyer and Allen, 2009; Commander et al., 2020; Gioria et al., 2021).

A functional definition of dormancy is when a viable seed fails to germinate when provided conditions under which a nondormant seed of the same species would be able to complete germination. Seed dormancy is generally a transient developmental state that eventually disappears to allow the seed to germinate. Dormancy alleviation can take place in one of two common ways: through aging of a dry seed (called "after-ripening") or through extended exposure of imbibed seeds to cold temperatures (called "chilling" or "stratification"). Between a fully dormant state in which a seed will not germinate under any condition and a fully nondormant state in which the seed will germinate across a wide range of conditions, there are many intermediate states in which germination is conditional on the environment being within certain ranges (e.g., temperature or water potential) or providing certain signals (e.g., light). PBT models can therefore be useful for quantifying the status of a seed population in terms of the environmental thresholds (i.e., "niche") permitting germination at a given time, and how these thresholds change in response to dormancy-breaking conditions (Bradford and Bello, 2022).

Seed dormancy is also a good illustration of individual developmental behavior, as at a given time and under a specific condition, dormancy is defined as the fraction of the viable seeds that do not germinate. Changes in the depth of dormancy can be defined by the shifting of sensitivity threshold distributions toward more likely to germinate or less likely to germinate under specific environmental conditions, for example, temperature or water potential. As such, population-based hydrothermal time models have been effectively employed for both physiological and ecological analyses of seed dormancy (Alvarado and Bradford, 2005; Meyer and Allen, 2009; Batlla and Benech-Arnold, 2015). With respect to physiological aspects of dormancy, both the hydro and the thermal components can contribute. For example, a progressive shift in the water potential threshold distribution (toward more stress tolerance) is associated with the loss of seed dormancy due to after-ripening (Christensen et al., 1996; Bair et al., 2006), which enables more seeds to germinate and to do so more quickly. Dormancy is also controlled by shifts in the temperature thresholds for germination. Seeds of many species will germinate only within a limited temperature range when they are dormant, but this range can expand as dormancy is released (Fig. 7.4A) (Batlla and Benech-Arnold, 2015). These minimum (T_{lower} or T_l) and maximum temperature (T_{higher} or T_h) thresholds for the alleviation of dormancy can shift seasonally in seeds in the soil seed bank and in response to various dormancy-breaking conditions. In general, dormancy-breaking conditions tend to expand the temperature range to the limits seen in nondormant seeds (Fig. 7.4B). In contrast, conditions that induce dormancy in these species narrow the range of temperatures at which germination is permitted and can merge together and prevent germination at any temperature (Fig. 7.4A), as in the induction of secondary dormancy (Malavert et al., 2021b). This approach has been applied successfully to understand the emergence differences between summer annual and winter annual weeds and their responses to additional environmental inputs (Fernández Farnocchia et al., 2019; Fernández Farnocchia et al., 2021; Malavert et al., 2021a).

This PBT approach also was tested experimentally by Huang et al. (2016) in relation to the population dynamics of a desert plant community. Observations of populations of thirteen winter annual species in the Sonoran Desert over twenty-five years included detailed demographic data of germination dates, germination fractions, survival to reproduction, numbers of seeds produced, and so on (Venable, 2007). These data were compared to the germination traits of seeds collected and tested in one year using the PBT model to quantify their responses to temperature and water availability (water potential) (see Chapter 2; Bradford, 1990; Alvarado and Bradford, 2002). Some of these parameters, including threshold (base) water potentials and thermal times required for germination, were highly correlated with other measures of success in this environment (Huang et al., 2016). As predicted from bet-hedging theory (Gremer and Venable, 2014), some species having high year-to-year variance in seed production per germinant had lower germination fractions or greater dormancy.

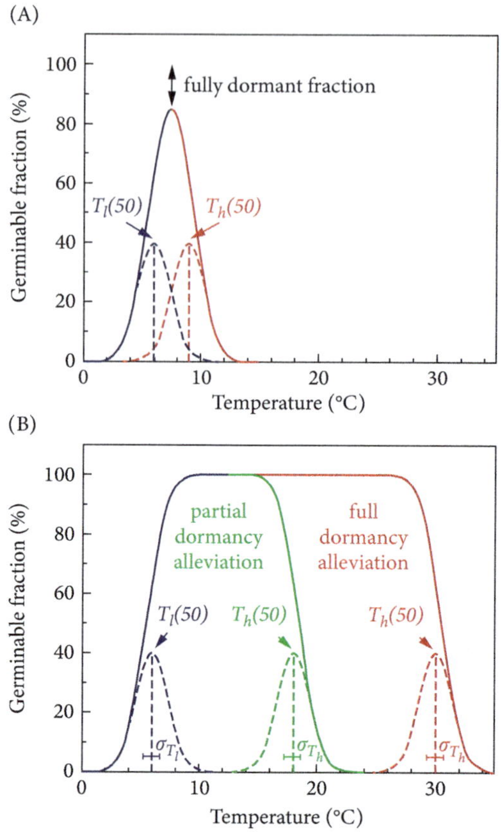

Figure 7.4 Temperature thresholds for the alleviation of seed dormancy in a winter annual species. **A.** A winter annual seed lot exhibiting dormancy due to a low high-temperature limit (T_h) for dormancy alleviation such that its distribution overlaps that of the lower temperature limit (T_l). This results in a narrow temperature range in which some seeds are above their T_l but still below their T_h, but most seeds remain dormant. $T_l(50)$ and $T_h(50)$ indicate the medians of the lower and higher temperature limit distributions at which dormancy can be alleviated. **B.** After being exposed to conditions within the range that alleviates dormancy (e.g., chilling or after-ripening), the T_h distribution shifts to higher temperatures (green and red distributions), enabling a greater fraction of the seeds to lose dormancy. The opposite occurs in summer annual species, with loss of dormancy occurring only at lower temperatures, and the T_l distribution shifts to lower temperatures in response to dormancy-alleviating conditions.

Adapted with permission (Open Access) from Bradford, K.J. and Bello, P. (2022). Applying population-based threshold models to quantify and improve seed quality attributes. In Buitink, J. and Leprince, O. (Eds.) Advances in Seed Science and Technology for More Sustainable Crop Production. Cambridge, UK, Burleigh Dodds Science Publishing. https://doi.org/10.19103/AS.2022.0105.05

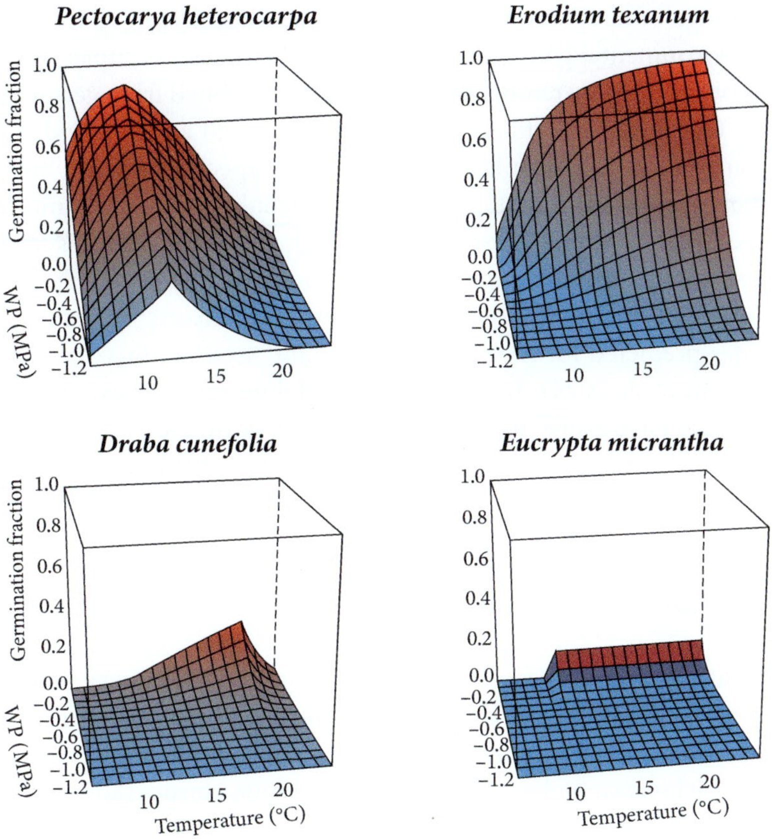

Figure 7.5 Simulated germination fractions within 5 days of incubation in a matrix of water potentials (WP) and temperatures for four desert annual species (illustrative of the four germination ecology syndromes identified by cluster analysis). Simulations were generated by the hydrothermal time model after 4 months (5.5 months for *Eucrypta micrantha*) of dry after-ripening in the field.

Reproduced with permission from Liu, S., et al. (2020). Hydrothermal sensitivities of seed populations underlie fluctuations of dormancy states in an annual plant community. *Ecology* 101(3): e02958. https://doi.org/10.1002/ecy.2958.

In a further study, Liu S. et al. (2020) examined the germination characteristics of these species as dormancy was lost due to after-ripening (dry storage). This study identified wide variation among species in their germination sensitivities to temperature and water potential, which were correlated with their mean germination date and final germination fraction in the field. The kinetics of dormancy loss during after-ripening also varied among species and were correlated with germination success in the field. The application of the PBT model to these data enabled prediction of germination responses across a wide range of temperatures and water potentials, that is, characterization of the environmental germination niche that the species occupies, with some species preferring low temperatures and being inhibited from germinating at warmer temperatures, while others germinated well across a wide temperature range (Fig. 7.5). Similarly, seeds of some species required high water potential for rapid germination, while others could germinate at much more negative water potential (greater stress). These response surfaces for germination sensitivity to temperature and water potential can be readily generated using the PBT model, enabling a mechanistic comparison of the ecological niches predicted for rapid germination and seedling success in the field (Rajapakshe et al., 2024). Similarly, PBT modeling approaches can be used to predict weed emergence timing based upon the dormancy-breaking requirements of seeds (Allen and Meyer, 2002; Boddy et al., 2012; Boddy et al., 2013; Batlla and Benech-Arnold, 2015; Hawkins et al., 2017; Malavert et al., 2021b).

Conclusions

Biological development, ecology and evolution are each very large topics, and I do not claim to have done them justice here. However, the examples cited are sufficient to indicate that the PBT concept deserves consideration as a comprehensive biological hypothesis. Despite considerable attention to the site(s) and role(s) of variation in organisms' responses and adaptations to their environment and how these find their way into genetic toolboxes, blind spots still exist. The PBT concept provides an organizing principle that can be applied across all relevant organizational levels and can be partitioned among genetic, physiological or developmental variation. The occurrence of bet-hedging, for example, elevates the role of individuals as the carriers of genetic information to the subsequent generation, while epigenetic markers incorporated in progeny during the reproductive process allow "preadaptive" variation to also play a role in fitness. While much of evolutionary biology has been conceived under a population genetics umbrella, genetics is not the only biological process that operates at a population level. In fact, as has been documented in this book, all biology can be conceived of as being built upon underlying populations at all levels of complexity (Fig. 2.7). This is a self-generating and self-regulating system that must have been present at the very beginning of life and that still underpins it in all of its astounding variety.

8

Implications and Anticipations

Individuality and population-based variation

The premise of this book is that a more complete understanding of biology requires an appreciation of the importance of individuality and its population-based variation at all levels of structure and complexity. If this premise is correct, then what are the implications for the conduct of biological science, and what might we anticipate as consequences if a population-based approach was adopted broadly? What changes would be needed in experimental design and interpretation to incorporate this viewpoint? How would statistical analyses need to be modified if the focus were no longer solely on means and significant difference probabilities? What conceptual paradigms would need to be adopted or rejected to take advantage of insights derived from the PBT viewpoint? Furthermore, if these questions are answered successfully and a PBT approach is adopted, what might we anticipate in the future of biological sciences? What novel information might we access, and how might it be applied? How do biological systems create and maintain both individuality and community? How are such population-based systems established, maintained and managed in cells and organisms and adapted through evolution? Finally, if the PBT approach is effective in enhancing our understanding of biological systems, to what extent might it also apply to other areas of interest, such as human social, economic or political systems? These are all large topics, so I will only be able to give my own views and suggestions. It will require experts in each aspect of biology (including psychology and sociology) to examine the fundamentals of their fields with a PBT viewpoint and see what can be found. To return to our consistent metaphor, what would we see if all biologists put on population-colored glasses and looked anew at their specialties?

Populations of Individuals. Kent J. Bradford, Oxford University Press. © Kent J. Bradford (2026).
DOI: 10.1093/9780197846032.003.0008

Implications

Methodological

The PBT approach has many direct implications for the methodology of biological science. Specifically, an emphasis on measuring individuals rather than bulk sampling and averaging is required. That is not to say that the latter methods were not or do not continue to be valuable. The technologies for characterizing fundamental individual biological entities were hardly conceptualized, much less realized, until recently. Nonetheless, a great deal was learned about how cells function from such pooled or tissue-level studies, but exploring deeper into biological systems requires an individualized approach. Fortunately, a plethora of new methods are available for both individual (single-cell) and spatial (localized) discrimination and characterization of cells at the chromatin, RNA, protein and metabolite levels, as referred to in earlier chapters. It is now possible to measure and characterize single-cell transcriptomes, proteomes or metabolomes, and to use this information to categorize cells into subpopulations having distinct patterns. Single-cell analyses enable categorization by population-based approaches, in contrast to the bulking together of hundreds of distinct cell types into pooled samples in which the critical differentiation among such cells is obscured. For example, a recent single-cell analysis of the entire brain of *Drosophila melanogaster* identified 8453 cell types within a total of 139,256 neurons (Schlegel et al., 2024). Interestingly, 59% of those cell types were catetorized as "singleton," or present as only a single member, the limit of individuality. At the same time, these characterizations can identify unique features of such subpopulations so that they can be separated and tested independently for their distinct (but still potentially overlapping) functions. However, those fields that were the first to adopt these approaches initially struggled to understand how to cope with that flood of diversity, for example, by simply calling it "noise" or "instability" and focusing on mechanisms for its suppression.

It is becoming clearer every day that the meaning and information in that "noise" can be accessed by using those same methods to pool together only individuals sharing certain defining characteristics that identify them as members of a related subpopulation. Studies of various organs and tissues in both plants and animals have revealed that they are composed of hundreds to thousands of differentiated subcategories of cells (despite sharing identical genomes), each doing its specific tasks to contribute to the overall viability and adaptability of the organism (Hu et al., 2024; Kanatani et al., 2024; Gandin et al., 2025). Once the cells are recognized as being of a certain subgroup, they can be identified based on those unique characteristics and then studied to understand what levels of which signals they respond to, enabling their threshold distributions and response rates to be assessed. Enzymology, for example, progressed from studying pooled tissue extracts that could convert a given substrate to a specific product, to associating those transforming entities with specific proteins, which are coded in specific genes but which also can assume a wide range of conformations and have dozens of chemical modifications, with complex feedback and feedforward loops at each of those gene, protein, conformation and modification levels. We can now appreciate the individuality of protein molecules and how they could be recruited in response to substrate concentrations or shut down in response to feedback signals or the cellular environment. This has implications for understanding some of the basic tools utilized in the field, such as Michaelis–Menten enzyme kinetics (see Fig. 3.6). The same analysis can be applied to cells with distinct developmental trajectories to fulfill specific biological functions or to the phenotypic variation among genetically identical individuals or near-identical plants or animals (species) that allows them to extend across ecological ranges.

The PBT approach helps view this complex web as coexisting in consistent subpopulations that have very specific tasks, but that can vary both in their responses to different signals or inputs and in the levels of these inputs that trigger them. Thus, while this approach is based upon individuals with some distinct properties, it also assumes that they can be grouped into communities represented by sensitivity distributions associated with their functions under specific environmental or physiological conditions. This approach is much different, and much more powerful, than attempting to just average out the variation or to pool subgroups together but only pay attention to the overall mean. The mean value remains significant as a characteristic of collective sensitivity, but means can also be quite misleading when members of the population vary widely in their sensitivities and effects, often over orders of magnitude in log-normal distributions (see, e.g., Figs. 4.5, 5.9, 6.7). Only a fraction of the population can be doing all the work at low factor concentrations, but we conceptualize that it is occurring in a graded manner in all members of the population, since they are all assumed to be identical. Instead, we will first need to define the populations (and subpopulations) and examine how the necessary tasks and sensitivities are distributed among them.

We also need some rethinking about signal strengths of inducing or repressing factors, which have often been analyzed in terms of movement by diffusion with consistent declines in signal concentration with distance from a source (e.g., Fig. 7.1). However, any signal/response system requires both the signal and the responding receptor (Bradford and Trewavas, 1994), and as illustrated repeatedly here, the receptors can also vary in their numbers, binding affinities, spatial localizations and downstream effects. Thus, the system is much more complex than initially envisioned based primarily on diffusion kinetics and rate constants and assuming a common binding affinity among identical receptor molecules. Certainly, physics still rules at very small distances, but the transitions in scale from the atomic to the biophysical to the physiological and cellular bring in biological capabilities and emergent properties that are not so constrained by those underlying physical principles (McFadden and Al-Khalili, 2014; Parthasarathy, 2022).

In summary, the primary methodological change that the PBT approach requires is to see the standing variation in biological entities and processes not as "merely" experimental error or "noise," but rather as the reality of consistent variation in sensitivities and functionality among the individual biological entities being measured or observed. New technologies have enabled studies of individual molecules and cells, and appropriate technologies are available at all levels of organization. Nonetheless, the concept of individuality that has thus been revealed (e.g., Trewavas, 1999) has yet to be fully accepted, appreciated and generalized. When we look at data that show a sigmoid range of activity or a distribution of response modes, we should recognize immediately that this likely represents actual individuals exhibiting the breadth of those characteristics, not just technical error around a "true" mean value. Each position within the threshold distribution represents a specific fraction of the receptor population, not just a statistical probability of "error" around a mean. The questions we must then ask are: what underlies these individual differences, how are these standing diverse populations created and maintained, and how do they collectively result in the observed biological behaviors? Until the existence of individual variation is accepted, these questions will not be asked, and therefore experiments will not be planned and executed to identify such variation rather than obscure and misinterpret it in pooled samples.

Statistical

The discussion of means and errors brings us to the topic of the statistics applied to biological data. I am not a statistician, and I am indebted particularly to Aubrey Clayton (2021) for providing a professional and technical basis for my own concerns about assumptions and constraints embedded in "standard" statistical methods (and I have had my share of arguments with adamant reviewers and critics over them). I refer the reader to Clayton's book to flesh out the background I am applying here to the current issue. A professional biologist, he maintains, is required to become adept in a specific set of statistical techniques whose purpose is to determine the "truth" or "reliability" or "probability" of data, or basically, the extent to which you and others can believe and trust your findings. Most statistical techniques broadly regarded as necessary for proper scientific analysis of data are based upon determining whether two means are "significantly" different. Even though most of these techniques require that the actual data represents a normal distribution for the techniques to be legitimately applied, in general, only the means are reported, and the variances around those means are used to assign levels of significance, or probability, that a given mean differs from another one (with its own variance). And if an expected or a desired difference was not detected, increasing the number of replicates almost always ensures that the required number of significance asterisks will appear, and the work will then be publishable. In fact, with enough replicates, any two non-identical means can be declared significantly different, even if it may be a minor "difference without a distinction."

Since the underlying normal distribution is assumed (or achieved by transformations of the data), most statistical approaches calculate a standard deviation, which quantifies the spread of the data around the mean. However, this standard deviation is then commonly divided by the square root of the number of replicates to give the standard error around the mean. For a given variance, increasing the number of replicates thereby reduces the standard error term around the mean. Note how the actual variance of the measured population, based on individual measurements, is thereby transformed into an error term, apparently due only to random variation in the measurements, not among the biological entities themselves. Measurement error certainly exists, particularly when common statistical procedures were being developed. However, as has been presented repeatedly in this book, we now can obtain individual measurements with quite small uncertainty in measurement, but we still refer to the variation among such individuals as the standard "error." This focus on methodological or experimental error (or just stochastic randomness) as the cause of biological variation represents a blind spot with respect to biological processes. Instead, the combination of the mean and the standard deviation more likely represents the actual midpoint, range, and frequencies of the individual biological variants under study (Fig. 2.2). Thus, a key change in statistical thought for biology would be to appreciate that the underlying variance distribution, rather than indicating the probability of error around a true mean, actually represents the relative frequencies of fractions of the population exhibiting different values of the feature being measured.

While making this criticism, I also recognize that changes are occurring in how data are presented. When many individual measurements are possible, as in single-cell analyses, it is becoming more common to use "violin plots" that show the actual data underlying the derived statistical values, often overlain by indications of the mean (and/or median) and range (e.g., 25 and 75%) of the values. These plots illustrate the presence of treatment effects on both the means and the distributions, but generally they only emphasize whether the means differ based on statistical probability calculations, which is virtually guaranteed given the very large number of individuals in the sample (e.g., transcripts, enzyme molecules, which can number in the thousands or millions). An example of data expressed in this way is shown in Figure 8.1 (Ouonkap et al., 2024). It is apparent that when the individual data points are presented, the simple comparison of means or medians is missing much of the story. While quantifying and comparing mean behavior is essential and valuable, there are also clear differences among the cultivars in the extent and shape of the variation among data points as well. Those changes may be equally interesting biologically, as these are in fact individuals being measured, and understanding how cellular populations exhibit such shifts in

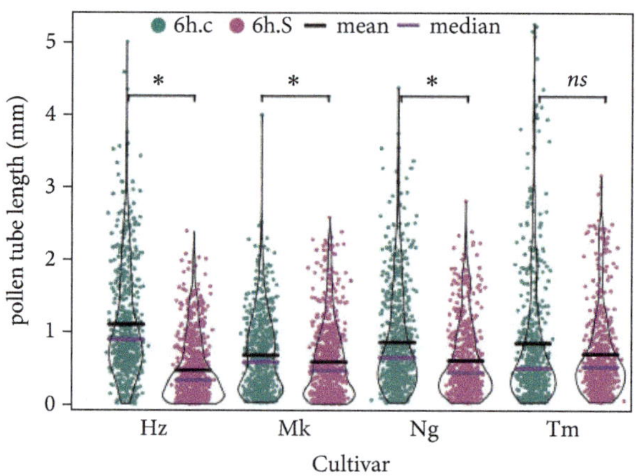

Figure 8.1 Measurements of pollen tube growth of different tomato cultivars under control (green) or heat-stressed (purple) conditions. This is shown solely as an example of current data presentation methods.

Reproduced with permission from Ouonkap, S.V.Y., et al. (2024). Enhanced pollen tube performance at high temperature contributes to thermotolerant fruit and seed production in tomato. *Current Biology* 34(22); 5319–5333. https://doi.org/10.1016/j.cub.2024.10.025.

sensitivity/response distributions is a key objective from a PBT-based viewpoint. For example, it is evident that the bulk of individuals share common values at or below a value of 1 and that reducing the range of variation is the primary effect of stress. This reduction in extreme values is reflected in the means and may justify calling them statistically significant, but this provides little information regarding mechanism. Alternatively, the behavior in Figure 8.1 could also be accounted for by specific stress effects on a subpopulation of individuals: the most rapidly growing individuals achieving the greatest lengths under control conditions exhibited the largest negative effects when stressed.

There are also clear examples of similar issues in the quantification of seed germination behavior. For example, how is one to quantify the speed of seed germination, as it varies among seeds in the population (see Fig. 5.8). Often, seed biologists use the time to 50% germination as a measurement for comparing seed lots. However, even this is not so simple (Bewley et al., 2013). In many cases, not all the seeds in a sample will germinate due to their being dormant or dead. Is the mean germination time then the time required for the 50th percentile of the total population to germinate, or the time until half of the seeds that can germinate have done so? Because these two approaches give very different answers (Soltani et al., 2015), clarity is needed in how the values are calculated. Other biological measurements (e.g., Michaelis–Menten kinetics) similarly use mean or median values in a population for characterizing rates (e.g., half of V_{max}; Fig. 3.6), often without considering or specifying the underlying basis for comparison (e.g., Fig. 4.4). If some of the hypotheses proposed here are valid, such as there being a reservoir of signaling proteins or receptors (as in cellular condensates) that can be drawn on as signal concentrations change, then the total active receptor population size can be changing as signal concentrations change. As a result, there will be different relative values for specific fractions of the population depending upon the factor level and the total population size. These considerations have seldom been appreciated in the past when all individuals were considered to be identical and equally accessible to a substrate or signal molecule, but a PBT-based viewpoint makes the importance of these issues apparent.

Conceptual

In the physical (non-biological) world, particles, atoms and molecules are all considered to be identical within their own category. Any variation must therefore be assigned either to inherent randomness (stochasticity) or to measurement error. It is assumed, for example, that at the scale at which it operates, diffusion randomly mixes molecules. However, once biology began, its overriding feature is the ability to thwart the thermodynamically expected increase in entropy, as is evident in the complexity of its structures and interactions. The game is different in biology, but it can neither ignore nor completely escape the underlying randomness. However, it can devise mechanisms to cope with it, and it can even turn stochasticity into a fundamental feature rather than a bug. If random variation is inevitable and is highly significant at the molecular level, how can biology live with it? One answer would be to incorporate the variation itself as a structural and organizing principle. That is what is embodied by the population-based approaches described in this book. The variation itself is harnessed as part of the mechanism, greatly simplifying the task of each individual unit in a population, which only needs to sense factors in its environment that are relevant to its function and respond in an on/off manner to perform the function it is programmed to do (Rodbard, 1973). The analog functions of biology are then accomplished by recruiting more or less of those individuals as needed to satisfy the required demand. Each chapter in this book has illustrated how this fundamental concept can be applied and confirmed in extant data across biology, from molecules to ecosystems. I note again that this book focuses on variation that occurs within and among individuals that either have a common genome (individual organisms) or are closely related genetically (within a species). Thus, genetic variation per se is not the primary source of such diversity, although it can contribute to it (see, e.g., gene duplication below).

The fundamental PBT methodology is to measure dose-response relationships, which can relate to many different inputs and applications (environmental factors, nutrients, pathogens, medicines, etc.), and then to develop models that allow the characterization and prediction of population-based responses to them. Thus, applying PBT models focuses our attention directly on the underlying mechanism that is most critical for practical application: the dose-response relationship between effectors and receptors. The models also define the variation in sensitivities that determines the range of doses needed to initiate or maximize the response of interest. Of course, dose-response relationships are characterized across biology, particularly in fields such as pharmacology. However, as noted previously (Chapter 3), that field's fundamental principles include an assumption that "the response observed is mediated by a uniform population of receptors" (Neubig et al., 2003, p. 601). The key conceptual introduction from the PBT model is that receptors vary in their sensitivities, likely through variations in their numbers, their conformations or their post-translational modifications. Thus, a simple dose-response curve becomes the method by which such a sensitivity distribution is revealed, identified and quantified.

The scale of sensitivity variation also needs to be given more attention when we are considering mechanisms. In many biochemical systems, a range of two orders of magnitude (often much larger) is commonly observed from the most sensitive to the maximal response. When we look at these sigmoid log-normal curves, we neglect to think seriously about what that range of variation implies. In plants at least, some hormonal responses cover up to six orders of magnitude (1,000,000-fold) in concentration to trigger the most sensitive to least sensitive responders (see, e.g., Fig. 5.8C). At the same time, if measurements of the amounts of these hormones detected in pooled tissue samples only double (with highly significant mean separations, of course), a trivial change in relation to the potential sensitivity range, this is taken as strong evidence of hormonal control of the process. How do we reconcile these two observations? In the case of seeds, for example, it has been shown that as few as 50 individual cells in the embryo may be responsible for the entire hormonal response system controlling germination of an individual seed (Topham et al., 2017). Thus, the changes in hormonal concentration might be orders of magnitude larger if all of the active hormone is concentrated into a tiny fraction of the total cell population rather than being evenly distributed in the tissue, as we assume when we pool many seeds and their diverse internal tissues together to extract their hormones. Laser microdissection and capture techniques have shown the wide variation in transcriptomes, for example, among cells in a developing plant embryo (Belmonte et al., 2013; Pelletier et al., 2025), indicating the distinct sensitivities and specific transcription factors active in such cells. The *in situ* spatial techniques that are now available will vastly expand such examples, making a population-based approach essential in understanding how such variation is managed to maintain both biological stability and adaptability.

Anticipations

Creation and maintenance of population variation

This book has presented many cases to support the contention that individuals at a given level of biology will be present in a distribution of sensitivities to factors influencing their function. The PBT approach enables characterization and quantification of the features of such populations, but it does not explain how they are created or maintained. While based in part on fundamental molecular randomness and an inability to completely control the conformations and activities of individual proteins, it is unlikely that biology would let such randomness run unchecked, just as it created complex editing and repair systems for DNA maintenance to counter random mutation. One approach, as illustrated in the olfactory and light-sensing systems of animals (Figs. 6.5, 6.6), is to distribute specific functions to distinct cells and maintain their individuality throughout their aggregation. However, the consistency with which such sensitivity threshold distributions can be identified at all levels of biological complexity suggests that they are fundamental features of biological systems. Thus, how sensitivity thresholds and their variation among individual units are created, monitored, adjusted or calibrated are key unknowns in biological function. However, if such variation is considered to be just experimental or methodological noise, we will not take it seriously as a biological feature, nor will we design experiments to uncover its basis. Bulked assays focusing on mean responses of pooled cell samples will never reveal underlying mechanisms that are based upon distributions of cell types and sensitivities. A key task going forward will be to understand how gene regulation (including epigenetics), transcription, translation, post-translational modification, spatial location, cell type variation, and so on, are managed to create standing populations of distinct individuals with the proper sensitivity distributions such that their integrated actions successfully carry out the essential functions of life. Very little of the individualized data available (e.g., single-cell transcriptomics, proteomics and metabolomics) has been analyzed from a PBT viewpoint, although underlying population distributions often can be readily seen in the data presented. The origins and management of such populations are therefore unknown, but some suggestions are discussed below.

Gene families and epigenetics

Functional genes (encoding, e.g., transcription factors, enzymes, receptors) are commonly present in multiple copies in the genome. This is generally thought to be due to duplication and subsequent accumulation of nucleotide variation in the encoding genes. Such expansion results in gene families that can specialize their expression in distinct cellular types and developmental phases. Polyploidy, in which multiple entire genomes are present in a species, has a similar effect in increasing opportunities for distinct expression of many gene family members (Morris et al., 2024). A common methological approach is to knock out members of such gene families systematically to identify their individual characteristics. It is not uncommon that loss of a single gene will have no discernible phenotypic consequence, a case of genetic redundancy (Nowak et al., 1997). Knocking out more members of the gene family and accumulating them in the same individual organism often eventually causes a phenotypic change that can be associated with the function of the gene family. This phenomenon has been examined largely from a genetic viewpoint focused on the implications for inheritance, breeding, evolution, and so on. For example, in the single plant genus *Solanum* (containing tomatoes, peppers and eggplants), "Despite broad conservation of gene macrosynteny among chromosome-scale references for 22 species, including 13 indigenous crops, thousands of gene duplications, particularly within key domestication gene families, exhibited dynamic trajectories in sequence, expression and function" (Benoit et al., 2025, p. 135). Expression of different, slightly varying genes encoding the same underlying function would result in a variable population of proteins associated with a common function, each potentially having a somewhat different sensitivity threshold for its transcription, substrates, signal molecules, and binding partners, among other things. The result would be overlapping subpopulations that would be difficult to distinguish using only the common pooled reaction or output. The challenge will be to deconvolute these overall response patterns into their constituents using individualized approaches (e.g., Figs. 6.1, 6.8), rather than relying on pooled samples and means. The same analysis applies to epigenetic mechanisms of transcriptional control, which in effect activate or silence genes both in real time and potentially extending into subsequent generations, creating the opportunity for mixed subpopulations from different gene groups and differential regulation of their relative compositions, particularly in relation to large-scale developmental changes (Marand et al., 2023; Engelhorn et al., 2025). PBT models can be helpful in identifying potential subpopulations able to be further subdivided (such as by fluorescent markers) and assayed for the specific genes expressed in each subpopulation.

Protein conformational variation

Even individual protein molecules apparently have different conformational options that can affect their activities (Liebherr et al., 2014; Fig. 3.5). We tend to forget that formerly, our knowledge of protein structures came primarily from X-ray crystallographic studies, where only a specific conformation is present (Myers, 2015). Protein molecules in an aqueous cellular environment have much more structural latitude. Computational approaches, such as AlphaFold3, can now predict many 3-D structures that are possible from the same amino acid sequence. This conformational variation itself could be the basis of sensitivity or enzymic efficiency variation among a population of proteins sharing the same amino acid sequences. My very preliminary mimicking of the Michaelis–Menten equation using the PBT model (Fig. 3.6) suggests that this variation may be logarithmic, implying that it is much broader than is generally appreciated. Furthermore, minor sequence variants among members of a gene family, as discussed above, can also have structural and functional effects. And after translation, "protein shapes are modulated by a vast array of small-molecule ligands and posttranslational modifications" (Qiao, 2025, p. 244). Thus, it may even be difficult for biological systems to produce the identical functional units that are generally assumed to be present, particularly considering the variable conditions in the subcellular environment (e.g., pH, mineral ions, osmotic pressure) that can affect the protein 3-D structure. More likely, this inherent variation was harnessed by emphasizing bimodal (quantal, digital) responses of diverse individuals in populations rather than analog ones requiring identical receptors that must monitor numerous input signals and respond quantitatively (Kaur et al., 2025). The increasing ability of methods based on artificial intelligence (AI) to predict protein structural variants will be valuable in quantifying and understanding the potential functional variation among these diverse structures (Escobedo et al., 2025).

Condensates

Condensates, or membrane-less cellular bodies resulting from liquid–liquid phase separation in cellular compartments, have been mentioned repeatedly in this book. They have properties that make them interesting in connection with the PBT model. For example, they can directly affect the variation in cytoplasmic protein content (Klosin et al., 2020). Klosin et al. termed such variation "noise," by which they meant inherent or stochastic variation in protein content due to differential transcription, translation and degradation rates. They also showed that thresholds in the cytoplasmic concentration of specific proteins can trigger condensate formation when exceeded, thereby stabilizing maximum cytoplasmic concentrations and reducing variation among individual cells (Fig. 8.2). In addition, specific amino acid sequences in proteins can influence their segregation and accumulation into specific

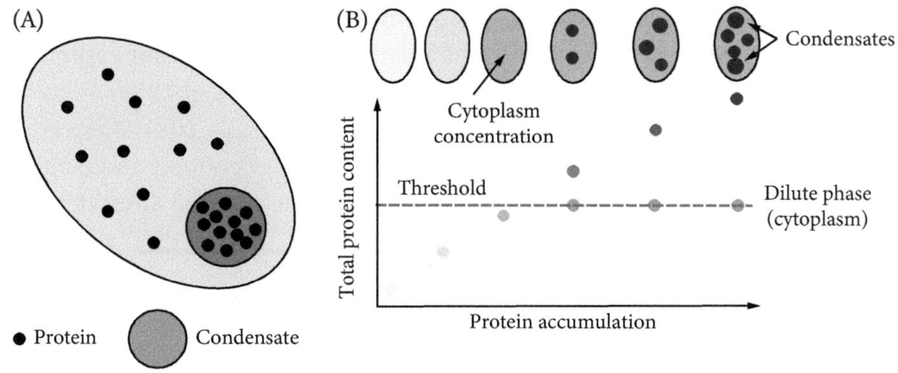

Figure 8.2 Condensates (liquid–liquid phase separations) can form when the concentration of proteins capable of trigging them exceeds a threshold concentration in the cytoplasm. **A.** Proteins are shown as dots, which are more diluted in the cytoplasm than in the condensates. **B.** As proteins accumulate, cytoplasmic concentrations increase to a threshold that triggers condensate formation, and subsequent protein accumulation occurs in the condensates, stabilizing cytoplasmic protein concentration among cells and moderating lack of synchrony between protein synthesis and degradation. This reservoir of proteins would fit well with the recruitment component of the PBT model, as when conditions change, the condensation threshold could shift, either sequestering or releasing the proteins in response to substrates, signaling factors or cellular conditions (pH, osmotic potential, etc.).

Adapted with permission from Klosin, A., et al. (2020). Phase separation provides a mechanism to reduce noise in cells. *Science* 367(6476): 464–468. http://doi.org/10.1126/science.aav6691

condensates within cells, indicating "that protein sequences contain not only a folding code but also a previously unrecognized code governing their distribution to diverse subcellular compartments" (Kilgore et al., 2025, p. 1095). From the PBT viewpoint, such cellular compartments provide convenient reservoirs for functional proteins that exhibit sensitivity thresholds in their behavior. We can imagine that the least sensitive members of the protein population might have a greater tendency to join condensates, due to variations in their amino acid sequences or posttranslational modifications. More sensitive individuals could remain in the liquid phase performing their designated function. However, if the concentration of their signal factor increases, exceeding the response capacity of the cytoplasmic members of the population, additional proteins could be released from the condensates to restore homeostasis (see, e.g., the discussion of blood glucose regulation in Chapter 6). Dai et al. (2023) demonstrated methods that can be used to isolate separately the contents of the condensates and of the cytoplasm, enabling experimental tests of this hypothesis. They also demonstrated that sequestration of charged ions also occurs in condensates, changing the electrochemical status of the cell and potentially triggering broader responses. That condensates can also adjust cellular osmotic potential (Watson et al., 2023) may be of particular interest in plants (Dorone et al., 2021; Field et al., 2023), which undergo much wider swings in cell water content and osmotic potentials than do most animal cells. This area is one of rapid advancement, and exploring mechanisms of condensate biology from a PBT viewpoint would lead to additional hypotheses and experimental approaches.

Analytical methods and artificial intelligence

Appropriate analytical methods are needed to characterize complete distributions of individual responses, rather than focusing only on comparing mean values using assumed underlying probability assumptions that lack appreciation of the multilayered complexity associated with biological processes. As noted above, the universal use of terms such as *standard error* reveals the basic assumption that variation around the mean is error, when in fact most of such variation likely represents the actual variability in biological systems. The mean is the average response among the members of that distribution, but it says nothing about the range of variation around it. In addition, while I have focused on the normal or Gaussian distribution, I am aware that many other distributions can be applied to fit different patterns of variation. I have no objection to these, except that in many cases skewness or lack of fit to a single normal distribution can be accounted for by a logarithmic rather than linear distribution scale or by the presence of multiple subpopulations that are (often unknowingly) being pooled in the dataset. Analytical methods focusing on identifying such subpopulations are urgently needed to apply to the enormous single-cell, single-nucleus, single-molecule datasets that are now being generated (Serrano et al., 2024). Using a population-based approach would focus more on clustering algorithms (e.g., Michelon et al., 2025) than on averaging to not simply separate such data by expression levels (which have less to do with physiology due to multiple layers of subsequent regulatory interventions) but rather to look for patterns in which subpopulations share key characteristics but not necessarily similar quantitative expression levels. Grouping such individuals into populations with variable expression levels of a similar cluster of genes could be more informative than current analysis techniques.

Artificial intelligence (AI) was mentioned only briefly in Chapter 3 in connection with the leap in protein structure prediction enabled by AlphaFold3. However, application of AI to biology is clearly here and will enable incredible advances. For example, an AI program called Evo 2 has been trained on procaryotic data to enable it "to capture two fundamental aspects of biology: the multimodality of the central dogma and the multiscale nature of evolution. The central dogma integrates DNA, RNA, and proteins with a unified code and predictable information flow, whereas evolution unifies the vastly different length scales of biological function represented by molecules, pathways, cells, and organisms" (Nguyen et al., 2024, p. 1). The TranscriptFormer single-cell model encompasses transcripts spanning 1.53 billion years of evolution across twelve species, "enabling it to function as a virtual instrument for probing cellular biology. . . . TranscriptFormer can be prompted to predict cell type-specific transcription factors and gene-gene interactions that align with independent experimental observations" (Pearce et al., 2025, p. 1). Similarly, the joint analysis of hosts and their associated microbial cells (termed "holobionts") is progressing rapidly. The functioning of plants and animals is dependent in multiple ways on their associated microbiota, such as nitrogen fixation from the air for plants or digestion of food for animals. Analyzing these together in the context of overall function greatly increases the scope of variation that could be operative in an organism. Due to the frequency of such distributions in biological data, computer programs that can readily identify and cluster individuals into related groups could be utilized to screen large datasets to identify subpopulations that underlie pooled data to produce the overall outcomes (Jacques et al., 2021). If I can do this empirically with a simple graphing program (Figs. 3.2, 4.5, 6.1, 6.8), imagine what AI could do if trained for this task.

At the same time, AI approaches can also have a blind spot with respect to recognizing the value of individuality. As Finn (2025, p. 479) noted, "The global competition to develop new forms of artificial intelligence (AI) frequently devolves into a race to see who can dehumanize people the fastest. This occurs when algorithms abstract away the individuality and context of billions of people whose creative outputs and personal information have been fed into large language models and other machine learning platforms—usually without their informed consent. The narratives embedded in many AI products also brush aside the intrinsic value of the individual by suggesting, directly or indirectly, that the amalgamated average is superior to the messiness of specific humans." In contrast, modifying AI algorithms to identify and cluster structured subpopulations among individualized data could respect individuality while also highlighting community.

Societal applications

This book has focused on biology, and primarily on how it functions collectively based on the contributions of diverse individuals at every level of complexity. Thus, to borrow again from Watson and Crick (1953), "it has not escaped our notice" that the PBT theory could also have applications in psychology and sociology with respect to the functioning of animal and human social systems, the most complex of biological systems. For example, the ability to recognize distinct individuals within a population enables social networks and impacts such actions as territoriality, competition, offspring identification, and mate recognition (Gokcekus et al., 2021). The complex social networks of bees and ants would also be ripe for PBT analyses to connect physiological and behavioral phenomena (Frank and Kronauer, 2024; Luo and Zhou, 2025; Sands et al., 2026). In a critique of sociobiology, Brown et al. (2025, p. 581) noted that "[e]volutionary models have . . . been applied to human cultural traits with the goal of understanding how the frequencies of social norms and institutions change over time. Through these efforts, we have learned that human cultural activities can generate niche construction, thereby influencing the form of natural selection acting back on a population and shaping developmental conditions for their offspring." Similarly, the PBT approach could also be extended to analyses of human behavior and social systems. As an example, elections are now accompanied by masses of polling data attempting to characterize and predict the preferences of voters. Such polling data frequently include collection of information on the responding individuals, such as age, race, gender, educational level and income, which are then used to identify subgroups and correlate them with their voting preferences. Different groups of humans will have different sensitivity thresholds for various social issues, such as the economy, immigration, personal freedom, taxation and international affairs, as can be identified by correlating the group affinities with voting behavior. In the end, only the sum of all those individual binary voting decisions determines the outcome of the election, but those decisions are determined by how individual voters ranked the importance of various factors to them to determine their choices. In addition, a given individual participates simultaneously in multiple subpopulations (e.g., being in both an age and an income category), and their sensitivities can vary among issues. For some, a key issue with high sensitivity can be the deciding factor, regardless of other issues; for others, the overall balance among sensitivities to several issues might be the determining factor. From a PBT perspective, the analysis of polling data amounts to characterizing the population-based thresholds of different subgroups to diverse issues in order to understand the influences of various issues and voter subgroups on the outcomes of elections.

Economics is another area where some aspects of PBT theory are being applied. Zilberman et al. (2012, p. 29) reviewed a "threshold model of adoption" of new technologies, which "assumes that individuals make adoption decisions using economic decision-making rules, heterogeneity of potential adopters, and dynamic processes—all factors that affect change over time." This model has much in common with the PBT model in relation to the variation among individuals in their adoption thresholds and the kinetics of "diffusion" within a population over time. As another example, DeMartino (2019) has discussed how economic projections need to consider the possible cost/benefit outcomes of a policy decision across a broad distribution of income levels. Policy outcomes are discussed primarily in a "cost/benefit" context, with the potential benefits often limited to those in one part of the distribution and relatively less attention paid to the accompanying negative effects that may be predicted for the opposite end of the distribution, which DeMartino terms the "economy harm profile." Similarly, the social costs of carbon generation (or of systems to reduce it) can differ depending upon a person's income (Prest et al., 2024). Such "distributional" or "equity" weighting can radically shift the cost/benefit balance of anticipated consequences. Variation in moral values among human individuals and populations must also be considered in economic analyses of human behavior, as noted by Landier and Thesmar (2025, p. 165):

> The main message of our book is a call for pluralism: it emphasizes the need to understand and incorporate diverse perspectives on moral-economic trade-offs. This pluralistic approach acknowledges that different persons may prioritize different, yet equally legitimate, values, such as freedom, justice, tradition, or economic efficiency.

I am not a sociologist, so I will refrain from trying to anticipate further PBT interpretations of human social systems, but from an observer's point of view (through my population-colored glasses) there seem to be many potential applications. I hope that some experts will be intrigued enough to give it a try. I will close with some thoughts from Barry Lopez (Lopez, 2022, p. 6):

> During my years of travel, my understanding of what diversity can mean has evolved. I began with an intuition, that the world was, from place to place and culture to culture, far more different than I had been led to believe. Later, I began to understand that to ignore these differences is not simply insensitive but unjust and

perilous. To ignore the differences does not make things better. It creates isolation, pain, fury, despair. Finally, I came to see something profound. Long-term, healthy patterns of social organization, among all social life-forms, it seemed to me, hinged on the work that maintained the integrity of the community while at the same time granting autonomy to its individuals. What made a society beautiful was some combination of autonomy and deference that, together, minimized strife. . . . In my understanding diversity is not, as I had once thought, a characteristic of life. It is, instead, a condition necessary *for* life. To eliminate diversity would be like eliminating carbon and expecting life to go on.

Conclusions

I commented in Chapter 1 that I am risking hubris to suggest that the PBT concept can be applied across all of biology. Having identified and discussed examples of such PBT behavior from molecules to ecosystems, I feel justified in at least raising the possibility that population-based variation in sensitivity thresholds underlies much of what transforms molecular entities into cells and cells into organisms. This conceptual change, allowing individuals within a given functional grouping to have distinct characteristics, rather than assuming that they are identical, might qualify as a Kuhnian paradigm shift in biology (Kuhn, 1962). If Kuhn is correct that we "see" reality through the lenses of our scientific paradigms, then I am recommending that biologists put on their population-colored glasses and look again at their own disciplines. We would not want to be like those who refused to even look through Galileo's telescope because what they might see would be incompatible with their long-held beliefs (Brecht, 1952). Our current ability to collect data at the individual level, whether on molecules, cells, organisms or people, is rapidly changing the underpinnings of biology. Biologists need to adopt a new perspective to fully understand and appreciate what these data are telling us. I urge you to put on your population-colored glasses and look again.

APPENDIX

Methods

Although the population-based threshold (PBT) model has a mathematical underpinning, this aspect was minimized in the text, as the book is intended for a broad audience of biologists, not mathematical modelers per se. In fact, I do not consider myself to be either a mathematician or a modeler, although I have sufficient familiarity with these fields to understand and use those aspects of the PBT approach. At the same time, I am encouraging fellow biologists to apply the PBT model in their own fields, and doing so requires more explanation about the underlying equations and methods. As the PBT model was first proposed and developed to describe seed germination behavior in response to temperature and water potential (Gummerson, 1986; Bradford, 1990; Dahal and Bradford, 1994), it is most widely referred to as a "hydrothermal time" model. However, it was subsequently expanded to describe many other aspects of seed biology, including respiration rates, dormancy, longevity and responses to hormones, oxygen, light, and other factors. These are all reviewed and described more thoroughly and mathematically in Bradford and Bello (2022), which the reader can consult for detailed information. In addition, broader aspects of the PBT approach were first described by Bradford and Trewavas (1994) and more recently by Bradford (2018) and Bradford and Bello (2023). This Appendix provides some of the key equations describing the model and some discussion of how to design experiments and analyze them using the PBT approach.

Dose-response relationships

The fundamental basis of the PBT model is the concept of thresholds for responses of individual biological entities (genes, proteins, cells, tissues, organs, organisms and communities) to signaling molecules or environmental conditions. As is common across biology, response thresholds are measured by dose-response relationships. Thus, the first step in applying the PBT model is to design and conduct experiments measuring the response output of the responding entities relative to the dosage of the inducing or controlling factor, preferably including the time course of the responses. The Michaelis–Menten relationship between the activity (substrate turnover rate) of an enzyme and substrate concentration is an example of such a dose-response experiment (see Fig. 3.6). For cell division, it is the time to division of a specific fraction of cells in response to a range of concentrations of a stimulatory factor (Fig. 6.3). In the case of seed germination, it is the speed and percentage of radicle emergence in relation to an environmental or physiological factor (Fig. 5.8). As the stimulatory factor concentration or intensity increases, an increasing fraction of the receptor population is recruited to respond. In addition, the speed of response of each recruited individual also increases in proportion to the amount by which the stimulus concentration (or level) exceeds the threshold of that individual. Inversely, as the concentration of an inhibitory factor increases, the time to response is lengthened proportionately, and an increasing fraction of the responsive population is inhibited from acting. Another fundamental tenet of the PBT model is that these thresholds vary in a consistent pattern (e.g., a normal distribution) among the individuals in the population. Putting these concepts together, we find that the basic PBT equation is:

$$\theta_X = (X - X_b(i))\, t_i \tag{A1}$$

where θ_X is the time constant for factor X, X is the concentration of factor X, $X_b(i)$ denotes the distribution of thresholds (base levels of factor X to trigger the receptors) among individual responders (i), and t_i is the time to response of individual i. Since θ_X is a constant, this shows that as the dosage (X) increases above the threshold of each individual [$X_b(i)$], their times to respond will decrease in inverse proportion (i.e., t_i becomes shorter; see Fig. 2.4). Alternatively, Equation A1 can be expressed as a response rate, or the inverse of the time to the event:

$$Response\ rate = 1/t_i = (X - X_b(i))\,/\theta_X \tag{A2}$$

The sensitivity threshold distribution $X_b(i)$ is described by a normal distribution across dosages or concentrations, which requires only its mean (μ) and standard deviation (σ) (see Fig. 2.2):

$$y = \frac{1}{\sigma\sqrt{2\pi}} e^{-\frac{(x-\mu)^2}{2\sigma^2}} \tag{A3}$$

The standard deviation is calculated by summing the differences of the trait values of each individual (x) from the overall mean, squaring it, dividing that by the number of individuals in the population, and then taking the square root, or:

$$\sigma = \sqrt{\frac{\sum(x_i - \mu)^2}{N}} \tag{A4}$$

where x_i is an individual value, μ is the mean and N is the total number of individuals. In practice, most spreadsheets and graphing programs have functions for calculating means and standard deviations and plotting normal distributions and their cumulative forms based on these parameters, so it is seldom required to use these equations directly.

The relationships in Equation A2 can be illustrated in a graph of response rates (RR) for different fractions of the responding population to different factor levels or dosages (Fig. A1). The corresponding time courses (on a linear scale) are shown in Figure A1B (see also Fig. 2.4). This figure shows that the RR values for different fractions increase in parallel above the threshold factor level for each fraction. However, actual response times increase in inverse proportion, with the least sensitive fractions being much more affected by decreases in the (stimulatory) factor. Inhibitors show the same curves, but with concentrations increasing in the opposite direction, which can be achieved by simply reversing the terms in parentheses in Equation A2:

$$Inhibitor\ Response\ Rate = 1/t_i = (X_b(i) - X)\,/\theta_X \tag{A5}$$

Note also that the threshold distribution $X_b(i)$ can be normal on a linear or a logarithmic scale (see, e.g., Figs. 2.4 and 3.7).

Fitting the PBT model to data

When originally proposed, the PBT model was fit to data using the approach shown in Figure A.1A, plotting the rates (inverses of times to germination) for specific fractions (percentages), determining their base values (intercepts on the x-axis), and plotting these versus the probit-transformed value of the fraction (Gummerson, 1986). Probits transform percentage data into "probability units" that are based upon the standard deviation of the normal distribution (Ellis and Roberts, 1981). Probits convert a sigmoid cumulative normal curve

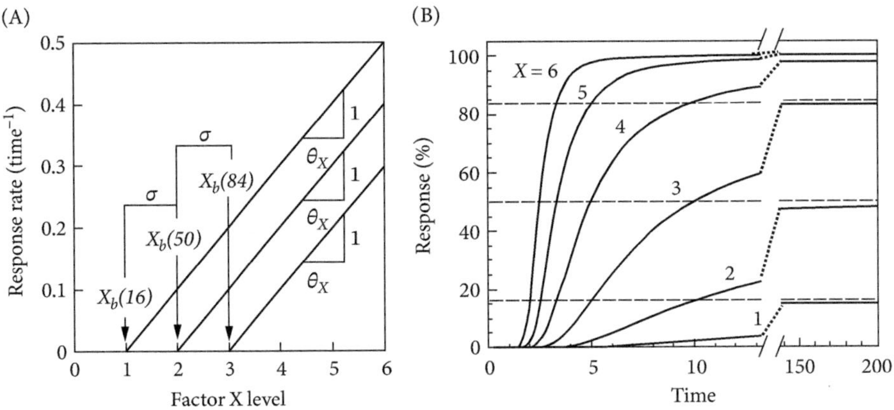

Figure A1 Response rates and time courses for the PBT model across factor levels. **A.** Response rates (inverses of times to respond, or $1/t_i$) as a function of stimulatory factor levels. Three lines are shown, representing the 16th, 50th and 84th percentage fractions of the responding population (or the median minus and plus one standard deviation). The intercepts on the factor axis represent the base (or threshold) factor levels for each fraction [$X_b(16), X_b(50), X_b(84)$]. The slopes of the lines are parallel and equal to the inverse of the time constant [$1/\theta_X$] (Eqn. A2). **B.** Time courses on a linear scale for responses at levels of factor X from 6 down to 1. This illustrates several features: (i) increasing (stimulatory) factor level *X* shortens the times to respond; inhibitory factor levels would increase in the opposite direction; (ii) the effect of changing factor levels is much more evident for the less sensitive fractions of the population (higher percentages; these are more sensitive to inhibitors); (iii) the break in the time axis illustrates that the time courses will approach the expected percentages (dashed horizontal lines at 16, 50 and 84%) after extended times. See also time course plots on logarithmic scales in Fig. 2.4C in the text.

into a straight line. However, this approach involved determining specific percentages from time course data, which required interpolating between actual data points. Bradford (1990) proposed instead plotting the probit of the observed percentage versus $X-(\theta_X/t_i)$, which is equal to $X_b(i)$ (a rearrangement of Eq. A1; see Eq. A6 below), across multiple factor levels. This results in linear relationships that converge to a single line when the optimal θ_X is identified. This convergence can be done by repeated probit regression analyses using different values of θ_X until the maximum fit (r^2) is identified (now done automatically in fitting programs). The $X_b(50)$ was then determined by the value at which the probit value was zero (50%) and the standard deviation (inverse of the slope) was one probit unit on either side of the median (probit = 0). This approach had the advantage that the observed rather than interpolated data could be used, and regression was easy to do in a spreadsheet.

There have been critics of this application of probit analysis (e.g., Onofri et al., 2014), and better methods can simultaneously fit all of the time courses at different doses of X to the basic equation. Equation A1 can also be expressed using the following equation:

$$\textit{Time to response for fraction } i = t_i = [X - (\theta_X/t_i) - X_b(50)]/\sigma \qquad \text{(A6)}$$

where $X_b(50)$ is the median and σ is the standard deviation of the normal distribution of $X_b(i)$. This equation can calculate the parameter values by simultaneously fitting response time courses across multiple factor levels (as shown in Fig. A1B). Once these parameter values are known, this equation can generate the time courses (t_i values) for any factor level. Optimization of this (using, e.g., Solver in Excel) provides the values of θ_X, $X_b(50)$, and σ (Bello and Bradford, 2016; Batlla and Agostinelli, 2017). Others have proposed alternative approaches, including maximum likelihood fitting (Moltchanova et al., 2020), and instructions in R coding language have been published for fitting probit-transformed percentage data (Wolkis et al., 2025); these can be adapted to fit the PBT model in Equation A6. An RShiny app has been created that embodies various seed-related applications of the PBT model, which could be expanded upon for other applications (P. Bello, B. Bradford, K. Bradford, unpublished; access through https://pbt-models.github.io/pbtm-site/). The key for any of these approaches is to fit more than one dosage or factor level simultaneously, thereby constraining the output to specific parameter values associated with the X factor dosages.

Subpopulations

It is often the case that the overall data represent the sum of two or more underlying subpopulations (e.g., Figs. 4.5, 5.12, 6.1, 6.8). These subpopulations can be fit and partitioned using an equation like the following:

$$T_r = f_1\,[[X - (\theta_{X1}/t_{i1}) - X_b(50)_1]/\sigma_1] + (1 - f_1)\,[[X - (\theta_{X2}/t_{i2}) - X_b(50)_2]/\sigma_2] \qquad \text{(A7)}$$

where T_r is the total response of the combined populations at different times, f_1 is the fraction of the total population in subfraction 1, and the subscript numbers refer to subpopulations 1 and 2 (see Fig. S7 in Bello and Bradford, 2016). The fitting process identifies the PBT parameters of each subpopulation and also optimizes the value of f_1, with the remainder of the population $(1 - f_1)$ being in the second population. Of course, this process could be extended to additional subpopulations if the data are sufficiently dense to support it. As is illustrated in Bello and Bradford (2016), closely spaced observations are required to separate subpopulations.

References Cited

Abley, K., et al. (2016) Developmental mechanisms underlying variable, invariant and plastic phenotypes. Ann. Bot. 117(5): 733–748. 10.1093/aob/mcw016

Abley, K., et al. (2021) An ABA-GA bistable switch can account for natural variation in the variability of Arabidopsis seed germination time. eLife 10(1): e59485. 10.7554/eLife.59485

Abramson, J., et al. (2024) Accurate structure prediction of biomolecular interactions with AlphaFold 3. Nature 630(8016): 493–500. 10.1038/s41586-024-07487-w

Adondakis, S., Venable, D.L. (2004) Dormancy and germination in a guild of Sonoran desert annuals. Ecology 85(9): 2582–2590. 10.1890/03-0587

Aebersold, R., et al. (2018) How many human proteoforms are there? Nature Chem. Biol. 14(3): 206–214. 10.1038/nchembio.2576

Aguirre Rivera, J., et al. (2024) Massively parallel analysis of single-molecule dynamics on next-generation sequencing chips. Science 385(6711): 892–898. 10.1126/science.adn5371

Alamos, S., et al. (2021) Quantitative imaging of RNA polymerase II activity in plants reveals the single-cell basis of tissue-wide transcriptional dynamics. Nature Plants 7(8): 1037–1049. 10.1038/s41477-021-00976-0

Albeck, J.G., et al. (2013) Frequency-modulated pulses of ERK activity transmit quantitative proliferation signals. Mol. Cell 49(2): 249–261. 10.1016/j.molcel.2012.11.002

Aldous, N., et al. (2023) Pancreatic β-cell heterogeneity in adult human islets and stem cell-derived islets. Cell. Mol. Life Sci. 80(6): 176. 10.1007/s00018-023-04815-7

Alfonso-Gonzalez, C., Hilgers, V. (2024) (Alternative) transcription start sites as regulators of RNA processing. Trends Cell Biol. 34(12): 1018–1028. 10.1016/j.tcb.2024.02.010

Allen, P.S., Meyer, S.E. (2002) Ecology and ecological genetics of seed dormancy in downy brome. Weed Sci. 50(2):241–247. 10.1614/0043-1745(2002)050[0241:EAEGOS]2.0.CO;2

Altschuler, S.J., Wu, L.F. (2010) Cellular heterogeneity: Do differences make a difference? Cell 141(4): 559–563. 10.1016/j.cell.2010.04.033

Alvarado, V., Bradford, K.J. (2002) A hydrothermal time model explains the cardinal temperatures for seed germination. Plant Cell Environ. 25(8): 1061–1069. 10.1046/j.1365-3040.2002.00894.x

Alvarado, V., Bradford, K.J. (2005) Hydrothermal time analysis of seed dormancy in true (botanical) potato seeds. Seed Sci. Res. 15(2): 77–88. 10.1079/SSR2005198

Ang, M.C.-Y., et al. (2024) Decoding early stress signaling waves in living plants using nanosensor multiplexing. Nature Comm. 15(1): 2943. 10.1038/s41467-024-47082-1

Angel, A., et al. (2011) A polycomb-based switch underlying quantitative epigenetic memory. Nature 476(7358): 105–108. 10.1038/nature10241

Angel, A., et al. (2015) Vernalizing cold is registered digitally at *FLC*. Proc. Natl. Acad. Sci. USA 112(13): 4146–4151. 10.1073/pnas.1503100112

Arias, A.M. (2023) The Master Builder: How the New Science of the Cell Is Rewriting the Story of Life. Basic Books, New York.

Assmann, S.M., et al. (2023) Rock, scissors, paper: How RNA structure informs function. Plant Cell 35(6): 1671–1717. 10.1093/plcell/koad026

Atanas, A.A., et al. (2023) Brain-wide representations of behavior spanning multiple timescales and states in *C. elegans*. Cell 186 (19): 4134–4151. 10.1016/j.cell.2023.07.035

Auge, G.A., et al. (2019) Pleiotropy in developmental regulation by flowering-pathway genes: is it an evolutionary constraint? New Phytol. 224(1): 55–70. 10.1111/nph.15901

Avrahami, D., et al. (2017) β-Cells are not uniform after all—Novel insights into molecular heterogeneity of insulin-secreting cells. Diabetes Obesity and Metab. 19(Suppl. 1): 147–152. 10.1111/dom.13019

Bair, N.B., et al. (2006) A hydrothermal after-ripening time model for seed dormancy loss in *Bromus tectorum* L. Seed Sci. Res. 16(1): 17–28. 10.1079/SSR2005237

Banani, S.F., et al. (2017) Biomolecular condensates: organizers of cellular biochemistry. Nature Rev. Molec. Cell Biol. 18(5): 285–298. 10.1038/nrm.2017.7

Barnard, E.A., et al. (1998) International Union of Pharmacology. XV. Subtypes of γ-aminobutyric acid A receptors: Classification on the basis of subunit structure and receptor function. Pharmacol. Rev. 50(2): 291–314.

Bartel, D.P. (2009) MicroRNAs: Target recognition and regulatory functions. Cell 136(2): 215–233. 10.1016/j.cell.2009.01.002

Baskin, C.C., Baskin, J.M. (2014) Seeds: Ecology, Biogeography, and Evolution of Dormancy and Germination, Second Edition. Elsevier Science, Burlington, VT.

Baskin, C.C., Baskin, J.M. (2020) Breaking seed dormancy during dry storage: A useful tool or major problem for successful restoration via direct seeding? Plants-Basel 9(5): 636. 10.3390/plants9050636

Baskin, J.M., Baskin, C.C. (2021) The great diversity in kinds of seed dormancy: a revision of the Nikolaeva–Baskin classification system for primary seed dormancy. Seed Sci. Res. 31(4): 249–277. 10.1017/S096025852100026X

Batlla, D., Agostinelli, A.M. (2017) Thermal regulation of secondary dormancy induction in *Polygonum aviculare* seeds: a quantitative analysis using the hydrotime model. Seed Sci. Res. 27(3): 231–242. 10.1017/S0960258517000198

Batlla, D., Benech-Arnold, R.L. (2015) A framework for the interpretation of temperature effects on dormancy and germination in seed populations showing dormancy. Seed Sci. Res. 25(Special Issue 02): 147–158. 10.1017/S0960258514000452

Bawa, G., et al. (2023) Introducing single cell stereo-sequencing technology to transform the plant transcriptome landscape. Trends Plant Sci. 29(2): 249–265. 10.1016/j.tplants.2023.10.002

Beal, J. (2017) Biochemical complexity drives log-normal variation in genetic expression. Engineer. Biol. 1(1): 55–60. 10.1049/enb.2017.0004

Bębenek, A., Ziuzia-Graczyk, I. (2018) Fidelity of DNA replication—a matter of proofreading. Curr. Genet. 64(5): 985–996. 10.1007/s00294-018-0820-1

Bello, P., Bradford, K.J. (2016) Single-seed oxygen consumption measurements and population-based threshold models link respiration and germination rates under diverse conditions. Seed Sci. Res. 26(3): 199–221. 10.1017/S0960258516000179

Bello, P.H.N. (2020) Analytical and Computational Tools to Assess Seed Quality and Model Germination Rates. MS Thesis. University of California, Davis, CA. https://www.proquest.com/dissertations-theses/analytical-computational-tools-assess-seed/docview/2540474371/se-2

Belmonte, M.F., et al. (2013) Comprehensive developmental profiles of gene activity in regions and subregions of the *Arabidopsis* seed. Proc. Natl. Acad. Sci. USA 110(5): E435–E444. 10.1073/pnas.1222061110

Benninger, R.K.P., Hodson, D.J. (2018) New understanding of β-cell heterogeneity and *in situ* islet function. Diabetes 67(4): 537–547. 10.2337/dbi17-0040

Benninger, R.K.P., Kravets, V. (2022) The physiological role of β-cell heterogeneity in pancreatic islet function. Nature Rev. Endocrin. 18(1): 9–22. 10.1038/s41574-021-00568-0

Benoit, M., et al. (2025) *Solanum* pan-genetics reveals paralogues as contingencies in crop engineering. Nature 640(1): 135–145. 10.1038/s41586-025-08619-6

Bentsink, L., et al. (2006) Cloning of *DOG1*, a quantitative trait locus controlling seed dormancy in Arabidopsis. Proc. Natl. Acad. Sci. USA 103(45): 17042–17047. 10.1073/pnas.0607877103

Bergson, H. (2016) Creative Evolution. Perennial Press, Moorestown, NJ

Bettenworth, V., et al. (2022) Frequency modulation of a bacterial quorum sensing response. Nature Comm. 13(1): 2772. 10.1038/s41467-022-30307-6

Bewley, J.D., et al. (2013) Seeds: Physiology of Development, Germination and Dormancy, Third Edition. Springer, New York.

Bierhuizen, J.F., Wagenvoort, W.A. (1974) Some aspects of seed germination in vegetables. I. The determination and application of heat sums and minimum temperature for germination. Scientia Hortic. 2(3): 213–219. 10.1016/0304-4238(74)90029-6

Bintu, L., et al. (2016) Dynamics of epigenetic regulation at the single-cell level. Science 351(6274): 720–724. 10.1126/science.aab2956

Blaby-Haas, C.E., Merchant, S.S. (2019) Comparative and functional algal genomics. Annu. Rev. Plant Biol. 70(1): 605–638. 10.1146/annurev-arplant-050718-095841

Blake, W.J., et al. (2006) Phenotypic consequences of promoter-mediated transcriptional noise. Mol. Cell 24(6): 853–865. 10.1016/j.molcel.2006.11.003

Bloom, A.J., et al. (1985) Resource limitation in plants-an economic analogy. Annu. Rev. Ecol. System. 16(1): 363–392. 10.1146/annurev.es.16.110185.002051

Boddy, L.G., et al. (2012) Population-based threshold models describe weed germination and emergence patterns across varying temperature, moisture and oxygen conditions. J. Appl. Ecol. 49(6): 1225–1236. 10.1111/j.1365-2664.2012.02206.x

Boddy, L.G., et al. (2013) Stratification requirements for seed dormancy alleviation in a wetland weed. PLoS ONE 8(9): e71457. 10.1371/journal.pone.0071457

Bohrer, C.H., Larson, D.R. (2021) The stochastic genome and its role in gene expression. Cold Spring Harbor Perspec. Biol.13(10): 1–20. 10.1101/cshperspect.a040386

Bradford, K.J. (1990) A water relations analysis of seed germination rates. Plant Physiol. 94(2): 840–849. 10.1104/pp.94.2.840

Bradford, K.J. (2018) Interpreting biological variation: seeds, populations and sensitivity thresholds. Seed Sci. Res. 28(3): 158–167. 10.1017/S0960258518000156

Bradford, K.J., Bello, P. (2022) Applying population-based threshold models to quantify and improve seed quality attributes. *In* J Buitink, and O Leprince, eds, Advances in Seed Science and Technology for More Sustainable Crop Production. Burleigh Dodds Science Publishing, Cambridge, UK, pp 1–88Cambridge, UK, pp 1–88. doi:10.19103/AS.2022.0105.05

Bradford, K.J., Bello, P. (2023) Understanding seed behavior: populations of individuals. Acta Hortic. 1365(1): 1–16. 10.17660/ActaHortic.2023.1365.1

Bradford, K.J., Hsiao, T.C. (1982) Physiological responses to moderate water stress. *In* OL Lange, PS Nobel, CB Osmond, H Ziegler, eds, Water relations and photosynthetic productivity, Vol. 12B, Encyclopedia of Plant Physiology, New Series. Springer-Verlag, Berlin, pp 263–324. doi:10.1007/978-3-642-68150-9_10

Bradford, K.J., Trewavas, A.J. (1994) Sensitivity thresholds and variable time scales in plant hormone action. Plant Physiol. 105(4): 1029–1036. 10.1104/pp.105.4.1029

Bradford, K.J., et al. (1993) A population-based threshold model describing the relationship between germination rates and seed deterioration. J. Exp. Bot. 44(264): 1225–1234. 10.1093/jxb/44.7.1225

Braitenberg, V. (1984) Vehicles: Experiments in Synthetic Psychology. MIT Press, Cambridge, MA.

Brecht, B. (1952) Galileo. Translated by Charles Laughton. Ed. Eric Bentley. Grove Press, 1966, New York.

Brehm-Stecher, B.F., Johnson, E.A. (2004) Single-cell microbiology: Tools, technologies, and applications. Microbiol. Mol. Biol. Rev. 68(3): 538–559. 10.1128/MMBR.68.3.538-559.2004

Bressan, D., et al. (2023) The dawn of spatial omics. Science 381(6657): eabq4964. 10.1126/science.abq4964

Bridson, E.Y., Gould, G.W. (2000) Quantal microbiology. Lett. Appl. Microbiol. 30(2): 95–98. 10.1046/j.1472-765x.2000.00673.x

Brixi, G., et al. (2025) Genome modeling and design across all domains of life with Evo 2. Arc Institute https://arcinstitute.org/manuscripts/Evo2

Brown, G.R., et al. (2025) Revisiting the human sociobiology debate. Science 389(6760): 580–581. 10.1126/science.ady6081

Bu, A., et al. (2024) Actuating extracellular matrices decouple the mechanical and biochemical effects of muscle contraction on motor neurons. Advanced Healthcare Materials 2403712(1): 1–18. 10.1002/adhm.202403712

Bueno, O., et al. (2018) Individuation, Process, and Scientific Practices. Oxford University Press, New York.

Burggren, W.W. (2019) Inadequacy of typical physiological experimental protocols for investigating consequences of stochastic weather events emerging from global warming. Am. J. Physiol.-Reg. Integ. Compar. Physiol. 316(4): R318–R322. 10.1152/ajpregu.00307.2018

Burggren, W.W., Mendez-Sanchez, J.F. (2023) "Bet hedging" against climate change in developing and adult animals: roles for stochastic gene expression, phenotypic plasticity, epigenetic inheritance and adaptation. Front. Physiol. 14(1245875): 1–19. 10.3389/fphys.2023.1245875

Burian, A., et al. (2022) Specification of leaf dorsiventrality via a prepatterned binary readout of a uniform auxin input. Nature Plants 8(3): 269–280. 10.1038/s41477-022-01111-3

Candela-Ferre, J., et al. (2024) Mind the gap: Epigenetic regulation of chromatin accessibility in plants. Plant Physiol. 194(4): 1998–2016. 10.1093/plphys/kiae024

Cao, D. (2022) To germinate or not? Transcriptional gradients underlie the seed dormancy continuum. Plant Physiol. 190(3): 1559–1561. 10.1093/plphys/kiac367

Capblancq, T., et al. (2020) Genomic prediction of (mal)adaptation across current and future climatic landscapes. Annu. Rev. Ecol. Evol. System. 51(1): 245–269. 10.1146/annurev-ecolsys-020720-042553

Cappell, S.D., et al. (2018) EMI1 switches from being a substrate to an inhibitor of APC/CCDH1 to start the cell cycle. Nature 558(7709): 313–317. 10.1038/s41586-018-0199-7

Chen, M., Chory, J. (2011) Phytochrome signaling mechanisms and the control of plant development. Trends Cell Biol. 21(11): 664–671. 10.1016/j.tcb.2011.07.002

Christensen, M., et al. (1996) A hydrothermal time model of seed after-ripening in *Bromus tectorum* L. Seed Sci. Res. 6(4): 155–164. 10.1017/S0960258500003214

Clark, I.C., et al. (2023) Microfluidics-free single-cell genomics with templated emulsification. Nat. Biotech. 41(1): 1557–1566. 10.1038/s41587-023-01685-z

Clayton, A. (2021) Bernoulli's Fallacy: Statistical Illogic and the Crisis of Modern Science. Columbia University Press, New York.

Cohen, J.D., Strader, L.C. (2024) An auxin research odyssey: 1989–2023. Plant Cell 36(5): 1410–1428. 10.1093/plcell/koae054

Commander, L.E., et al. (2020) Demographic, seed and microsite limitations to seedling recruitment in semi-arid mine site restoration. Plant Soil 457(1): 113–129. 10.1007/s11104-019-04081-2

Coombs, I.D., Farrant, M. (2023) An unexpected role for a glutamate receptor. Science 382(6677): 1363–1364. 10.1126/science.adm6771

Cortijo, S., Locke, J.C.W. (2020) Does gene expression noise play a functional role in plants? Trends Plant Sci. 25(10): 1041–1051. 10.1016/j.tplants.2020.04.017

Cortijo, S., et al. (2019) Widespread inter-individual gene expression variability in *Arabidopsis thaliana*. Molec. Syst. Biol. 15(1): e8591. 10.15252/msb.20188591

Cortijo, S., et al. (2020) Co-expression networks from gene expression variability between genetically identical seedlings can reveal novel regulatory relationships. Front. Plant Sci. 11(599464): 1–15. 10.3389/fpls.2020.599464

Dacey, D.M., et al. (2005) Melanopsin-expressing ganglion cells in primate retina signal colour and irradiance and project to the LGN. Nature 433(7027): 749–754. 10.1038/nature03387

Dahal, P., Bradford, K.J. (1994) Hydrothermal time analysis of tomato seed germination at suboptimal temperature and reduced water potential. Seed Sci. Res. 4(1): 71–80. 10.1017/S096025850000204X

Dahal, P., et al. (1990) Effects of priming and endosperm integrity on seed germination rates of tomato genotypes. I. Germination at suboptimal temperature. J. Exp. Bot. 41(232): 1431–1439. 10.1093/jxb/41.11.1431

Dai, Y., et al. (2023) Programmable synthetic biomolecular condensates for cellular control. Nature Chem. Biol. 19(1): 518–528. 10.1038/s41589-022-01252-8

Dai, Y., et al. (2024) Biomolecular condensates regulate cellular electrochemical equilibria. Cell 187(21): 5951–5966. 10.1016/j.cell.2024.08.018

Damasio, A. (1999) The Feeling of What Happens. Body and Emotion in the Making of Consciousness. Harcourt, Inc., New York.

Damer, B., Deamer, D. (2015) Coupled phases and combinatorial selection in fluctuating hydrothermal pools: A scenario to guide experimental approaches to the origin of cellular life. Life 5(1): 872–887. 10.3390/life5010872

Davies, A.E., et al. (2020) Systems-level properties of EGFR-RAS-ERK signaling amplify local signals to generate dynamic gene expression heterogeneity. Cell Syst. 11(2): 161–175. 10.1016/j.cels.2020.07.004

Davies, W.J. (1978) Some effects of abscisic acid and water stress on stomata of *Vicia faba* L. J. Exp. Bot. 29(1): 175–182. 10.1093/jxb/29.1.175

DeAngelis, D.L., Gross, L.J. (1992) Individual-based Models and Approaches in Ecology. Populations, Communities and Ecosystems. Chapman & Hall, New York.

Debat, V., David, P. (2001) Mapping phenotypes: canalization, plasticity and developmental stability. Trends Ecol. Evol. 16(10): 555–561. 10.1016/S0169-5347(01)02266-2

DeCuzzi, N.L., et al. (2024) Spatiotemporal clusters of ERK activity coordinate cytokine-induced inflammatory responses in human airway epithelial cells. Am. J. Respir. Cell Mol. Biol. 72(5): 520–532. 10.1165/rcmb.2024-0256OC

DeMartino, G.F. (2019) Econogenic harm and the case for 'economy harm profile' analysis. New Polit. Econ. 24(6): 798–815. 10.1080/13563467.2018.1526265

Dey, S.S., et al. (2015) Orthogonal control of expression mean and variance by epigenetic features at different genomic loci. Molec. Syst. Biol. 11(5): 806. 10.15252/msb.20145704

Dillon, M.E., et al. (2016) Life in the frequency domain: the biological impacts of changes in climate variability at multiple time scales. Integ. Compar. Biol. 56(1): 14–30. 10.1093/icb/icw024

Do, M.T.H. (2019) Melanopsin and the intrinsically photosensitive retinal ganglion cells: biophysics to behavior. Neuron 104(2): 205–226. 10.1016/j.neuron.2019.07.016

Dobzhansky, T. (1973) Nothing in biology makes sense except in the light of evolution. Am. Biol. Teacher 35(3): 125–129. 10.2307/4444260

Donohue, K., et al. (2015) Applying developmental threshold models to evolutionary ecology. Trends Ecol. Evol. 30(2): 66–77. 10.1016/j.tree.2014.11.008

Donohue, K., et al. (2010) Germination, postgermination adaptation, and species ecological ranges. Annu. Rev. Ecol. Evol. Syst. 41(1): 293–319. 10.1146/annurev-ecolsys-102209-144715

Dorone, Y., et al. (2021) A prion-like protein regulator of seed germination undergoes hydration-dependent phase separation. Cell 184(16): 4284–4298. 10.1016/j.cell.2021.06.009

Downton, W.J.S., et al. (1988) Non-uniform stomatal closure induced by water stress causes putative non-stomatal inhibition of photosynthesis. New Phytol. 110(4): 503–509. 10.1111/j.1469-8137.1988.tb00289.x

Drapek, C., et al. (2024) GA dynamics governing nodulation revealed using GIBBERELLIN PERCEPTION SENSOR 2 in *Medicago truncatula* lateral organs. Plant Cell 36(10): 4442–4456. 10.1093/plcell/koae201

Dror, E., et al. (2023) Epigenetic dosage identifies two major and functionally distinct β cell subtypes. Cell Metab. 35(5): 821–836. 10.1016/j.cmet.2023.03.008

Du, J., et al. (2024) Distinguishing individual photobodies using Oligopaints reveals thermo-sensitive and -insensitive phytochrome B condensation at distinct subnuclear locations. Nature Comm. 15(1): 3620. 10.1038/s41467-024-47789-1

Dudok, B., et al. (2024) Retrograde endocannabinoid signaling at inhibitory synapses *in vivo*. Science 383(6686): 967–970. 10.1126/science.adk3863

Dussutour, A. (2021) Learning in single cell organisms. Biochem. Biophys. Res. Commun. 564(1): 92–102. 10.1016/j.bbrc.2021.02.018

Ebert, M.S., Sharp, P.A. (2012) Roles for microRNAs in conferring robustness to biological processes. Cell 149(3): 515–524. 10.1016/j.cell.2012.04.005

Eckardt, N.A., et al. (2024) The lowdown on breakdown: Open questions in plant proteolysis. Plant Cell 36(9): 2931–2975. 10.1093/plcell/koae193

Eckert, L., et al. (2024) Biochemically plausible models of habituation for single-cell learning. Curr. Biol. 34(24): 5646–5658. 10.1016/j.cub.2024.10.041

Eisen, A.J., et al. (2024) Propofol anesthesia destabilizes neural dynamics across cortex. Neuron 112(16): 2799–2813. 10.1016/j.neuron.2024.06.011

Ellis, R.H. (2022) Seed ageing, survival and the improved seed viability equation; forty years on. Seed Sci. Technol. 50(2): 1–20. 10.15258/sst.2022.50.1.s.01

Ellis, R.H., Roberts, E.H. (1980) Improved equations for the prediction of seed longevity. Ann. Bot. 45(1): 13–30. 10.1093/oxfordjournals.aob.a085797

Ellis, R.H., Roberts, E.H. (1981) The quantification of aging and survival in orthodox seeds. Seed Sci. Technol. 9(2): 373–409.

Emani, P.S., et al. (2024) Single-cell genomics and regulatory networks for 388 human brains. Science 384(6698): eadi5199. 10.1126/science.adi5199

Emenecker, R.J., et al. (2021) Biological phase separation and biomolecular condensates in plants. Annu. Rev. Plant Biol. 72(1): 17–46. 10.1146/annurev-arplant-081720-015238

Emonet, T., et al. (2005) AgentCell: a digital single-cell assay for bacterial chemotaxis. Bioinformatics 21(11): 2714–2721. 10.1093/bioinformatics/bti391

Engelhorn, J., et al. (2025) Genetic variation at transcription factor binding sites largely explains phenotypic heritability in maize. Nature Genet. 57(1): 2313–2322. 10.1038/s41588-025-02246-7

English, B.P., et al. (2006) Ever-fluctuating single enzyme molecules: Michaelis-Menten equation revisited. Nature Chem. Biol. 2(2): 87–94. 10.1038/nchembio759

Esashi, Y., Leopold, A.C. (1968) Physical forces in dormancy and germination of *Xanthium* seeds 1. Plant Physiol. 43(6): 871–876. 10.1104/pp.43.6.871

Escobedo, A., et al. (2025) Genetics, energetics, and allostery in proteins with randomized cores and surfaces. Science 389(6758): eadq3948. 10.1126/science.adq3948

Exposito-Alonso, M. (2023) Understanding local plant extinctions before it is too late: bridging evolutionary genomics with global ecology. New Phytol. 237(6): 2005–2011. 10.1111/nph.18718

Fabrèges, D., et al. (2024) Temporal variability and cell mechanics control robustness in mammalian embryogenesis. Science 386(6718): eadh1145. 10.1126/science.adh1145

Falke, J.J., et al. (1997) The two-component signaling pathway of bacterial chemotaxis: a molecular view of signal transduction by receptors, kinases, and adaptation enzymes. Annu. Rev. Cell Dev. Biol. 13(1): 457–512. 10.1146/annurev.cellbio.13.1.457

Fan, M., et al. (2025) Reduced stomatal density in wheat overexpressing EPIDERMAL PATTERNING FACTOR1 differentially affects red and blue light responses. Plant Physiol. 199(1): kiaf379. 10.1093/plphys/kiaf379

Farack, L., et al. (2019) Transcriptional heterogeneity of beta cells in the intact pancreas. Devel. Cell 48(1): 115–125. 10.1016/j.devcel.2018.11.001

Farquhar, G.D., Sharkey, T.D. (1982) Stomatal conductance and photosynthesis. Annu. Rev. Plant Physiol. 33(1): 317–345.

Félix, M.-A., Barkoulas, M. (2015) Pervasive robustness in biological systems. Nature Rev. Genet. 16(8): 483–496. 10.1038/nrg3949

Fernández Farnocchia, R.B., et al. (2019) Regulation of seed dormancy by the maternal environment is instrumental for maximizing plant fitness in *Polygonum aviculare*. J. Exp. Bot. 70(18): 4793–4806. 10.1093/jxb/erz269

Fernández Farnocchia, R.B., et al. (2021) Optimization of timing of next-generation emergence in *Amaranthus hybridus* is determined via modulation of seed dormancy by the maternal environment. J. Exp. Bot. 72(12): 4283–4297. 10.1093/jxb/erab141

Ferrell, J.E., Machleder, E.M. (1998) The biochemical basis of an all-or-none cell fate switch in *Xenopus* oocytes. Science 280(5365): 895–898. 10.1126/science.280.5365.895

Ferri-Borgogno, S., et al. (2024) Molecular, metabolic, and subcellular mapping of the tumor immune microenvironment via 3D targeted and non-targeted multiplex multi-omics analyses. Cancers 16(5): 846. 10.3390/cancers16050846

Field, S., et al. (2023) Plants use molecular mechanisms mediated by biomolecular condensates to integrate environmental cues with development. Plant Cell 35(9): 3173–3186. 10.1093/plcell/koad062

Finch-Savage, W.E., Footitt, S. (2017) Seed dormancy cycling and the regulation of dormancy mechanisms to time germination in variable field environments. J. Exp. Bot. 68(4): 843–856. 10.1093/jxb/erw477

Finn, E. (2025) The monster in the room. Science 387(6733): 479–479. 10.1126/science.adu8851

Finn, E.H., Misteli, T. (2019) Molecular basis and biological function of variability in spatial genome organization. Science 365(6457): eaaw9498. 10.1126/science.aaw9498

Finn, E.H., et al. (2019) Extensive heterogeneity and intrinsic variation in spatial genome organization. Cell 176(6): 1502–1515. 10.1016/j.cell.2019.01.020

Frank, A., et al. (2024) The Blind Spot. Why Science Cannot Ignore Human Experience. The MIT Press, Cambridge, MA.

Frank, D.D., Kronauer, D.J.C. (2024) The budding neuroscience of ant social behavior. Annu. Rev. Neurosci. 47(1): 167–185. 10.1146/annurev-neuro-083023-102101

Franks, S.J., et al. (2014) Evolutionary and plastic responses to climate change in terrestrial plant populations. Evol. Appl. 7(1): 123–139. 10.1111/eva.12112

Friedman, R.C., et al. (2009) Most mammalian mRNAs are conserved targets of microRNAs. Genome Res. 19(1): 92–105. 10.1101/gr.082701.108

Fritz, M.A., et al. (2018) Mechanisms underlying the environmentally induced plasticity of leaf morphology. Front. Genet. 9(478): 1–25. 10.3389/fgene.2018.00478

Frontera, W.R., Ochala, J. (2015) Skeletal muscle: A brief review of structure and function. Calcif. Tissue Int. 96(3): 183–195. 10.1007/s00223-014-9915-y

Fuqua, W.C., et al. (1994) Quorum sensing in bacteria: the LuxR-LuxI family of cell density-responsive transcriptional regulators. J. Bacteriol. 176(2): 269–275.

Gandin, V., et al. (2025) Deep-tissue transcriptomics and subcellular imaging at high spatial resolution. Science 388(6744): eadq2084. 10.1126/science.adq2084

Gangestad, S.W., Thornhill, R. (1999) Individual differences in developmental precision and fluctuating asymmetry: a model and its implications. J. Evol. Biol. 12(2): 402–416. 10.1046/j.1420-9101.1999.00039.x

Gendron, J.M., Staiger, D. (2023) New horizons in plant photoperiodism. Annu. Rev. Plant Biol. 74(1): 481–509. 10.1146/annurev-arplant-070522-055628

Gianella, M., et al. (2021) Ecological, (epi)genetic and physiological aspects of bet-hedging in angiosperms. Plant Repro. 34(1): 21–36. 10.1007/s00497-020-00402-z

Gildor, T., Smadar, B.-T.d.-L. (2018) Comparative studies of gene expression kinetics: Methodologies and insights on development and evolution. Front. Genet. 9(339): 1–8. 10.3389/fgene.2018.00339

Gillies, T.E., et al. (2017) Linear integration of ERK activity predominates over persistence detection in Fra-1 regulation. Cell Syst. 5(6): 1–15. 10.1016/j.cels.2017.10.019

Gilroy, S., Trewavas, A. (2001) Signal processing and transduction in plant cells: the end of the beginning? Nature Rev. Mol. Cell Biol. 2(4): 307–314. 10.1038/35067109

Gilroy, S., Trewavas, T. (2023) Agency, teleonomy and signal transduction in plant systems. Biol. J. Linnean Soc. 139(4): 514–529. 10.1093/biolinnean/blac021

Gioria, M., et al. (2021) Persistent soil seed banks promote naturalisation and invasiveness in flowering plants. Ecol. Lett. 24(8): 1655–1667. 10.1111/ele.13783

Gokcekus, S., et al. (2021) Recognising the key role of individual recognition in social networks. Trends Ecol. Evol. 36(11): 1024–1035. 10.1016/j.tree.2021.06.009

Gotoh, E., et al. (2018) Chloroplast accumulation response enhances leaf photosynthesis and plant biomass production. Plant Physiol. 178(3): 1358–1369. 10.1104/pp.18.00484

Gremer, J.R., et al. (2020) Variation in the seasonal germination niche across an elevational gradient: the role of germination cueing in current and future climates. Am. J. Bot. 107(2): 350–363. 10.1002/ajb2.1425

Gremer, J.R., Venable, D.L. (2014) Bet hedging in desert winter annual plants: optimal germination strategies in a variable environment. Ecol. Lett. 17(3): 380–387. 10.1111/ele.12241

Gribaldo, S., Brochier-Armanet, C. (2006) The origin and evolution of Archaea: a state of the art. Philos. Trans. R. Soc. Lond. B. Biol. Sci. 361(1470): 1007–1022. 10.1098/rstb.2006.1841

Griffiths, J., et al. (2024) GIBBERELLIN PERCEPTION SENSOR 2 reveals genesis and role of cellular GA dynamics in light-regulated hypocotyl growth. Plant Cell 36(10): 4426–4441. 10.1093/plcell/koae198

Groot Crego, C., et al. (2024) CAM evolution is associated with gene family expansion in an explosive bromeliad radiation. Plant Cell 36(10): 4109–4131. 10.1093/plcell/koae130

Grujcic, V., et al. (2022) One cell at a time: Advances in single-cell methods and instrumentation for discovery in aquatic microbiology. Front. Microbiol. 13(881018): 1–10. 10.3389/fmicb.2022.881018

Gummerson, R.J. (1986) The effect of constant temperatures and osmotic potential on the germination of sugar beet. J. Exp. Bot. 37(179): 729–741. 10.1093/jxb/37.6.729

Guzikowski, N.J., Kavalali, E.T. (2024) Functional specificity of liquid-liquid phase separation at the synapse. Nature Comm. 15(1): 10103. 10.1038/s41467-024-54423-7

Haefner, J.W., et al. (1997) A spatially explicit model of patchy stomatal responses to humidity. Plant Cell Environ. 20(9): 1087–1097. 10.1046/j.1365-3040.1997.d01-137.x

Hallgrimsson, B., et al. (2019) The developmental-genetics of canalization. Semin. Cell Dev. Biol. 88(1): 67–79. 10.1016/j.semcdb.2018.05.019

Ham, L., et al. (2020) Extrinsic noise and heavy-tailed laws in gene expression. Phys. Rev. Lett. 124(10): 108101. 10.1103/PhysRevLett.124.108101

Ham, L., et al. (2021) Pathway dynamics can delineate the sources of transcriptional noise in gene expression. eLife 10(1): e69324. 10.7554/eLife.69324

Harton, M.D., Batchelor, E. (2017) Determining the limitations and benefits of noise in gene regulation and signal transduction through single cell, microscopy-based analysis. J. Mol. Biol. 429(8): 1143–1154. 10.1016/j.jmb.2017.03.007

Hawkins, K.K., et al. (2017) Secondary dormancy induction and release in *Bromus tectorum* seeds: the role of temperature, water potential and hydrothermal time. Seed Sci. Res. 27(1): 12–25. 10.1017/S0960258516000258

Herculano-Houzel, S. (2009) The human brain in numbers: a linearly scaled-up primate brain. Front. Human Neurosci. 3(31): 1–11. 10.3389/neuro.09.031.2009

Hermida, M. (2016) Life on Earth is an individual. Theory Biosci. 135(1): 37–44. 10.1007/s12064-016-0221-2

Herrera, C.M. (2009) Multiplicity in Unity: Plant Subindividual Variation and Interactions with Animals. University of Chicago Press, Chicago.

Herud-Sikimić, O., et al. (2021) A biosensor for the direct visualization of auxin. Nature 592(7856): 768–772. 10.1038/s41586-021-03425-2

Hesse, J.K., Tsao, D.Y. (2023) Functional modules for visual scene segmentation in macaque visual cortex. Proc. Natl. Acad. Sci. USA 120(32): e2221122120. 10.1073/pnas.2221122120

Heydenreich, F.M., et al. (2023) Molecular determinants of ligand efficacy and potency in GPCR signaling. Science 382(6677): eadh1859. 10.1126/science.adh1859

Hodge, R.D., et al. (2019) Conserved cell types with divergent features in human versus mouse cortex. Nature 573(7772): 61–68. 10.1038/s41586-019-1506-7

Hong, L., et al. (2018) Heterogeneity and robustness in plant morphogenesis: From cells to organs. Annu. Rev. Plant Biol. 69(1): 469–495. 10.1146/annurev-arplant-042817-040517

Hsu, P.Y., Harmer, S.L. (2012) Circadian phase has profound effects on differential expression analysis. PLoS ONE 7(11): e49853. 10.1371/journal.pone.0049853

Hu, Z., et al. (2024) Large-volume fully automated cell reconstruction generates a cell atlas of plant tissues. Plant Cell 36(12): 4840–4861. 10.1093/plcell/koae250

Huang, C.-H., et al. (2024) Editorial: Self-organizing and excitable signaling networks in cell biology. Front. Cell Devel. Biol. 12(1): 1430911. 10.3389/fcell.2024.1430911

Huang, Z., et al. (2016) The contribution of germination functional traits to population dynamics of a desert plant community. Ecology 97(1): 250–261. 10.1890/15-0744.1

Huo, H., et al. (2016) *DELAY OF GERMINATION1* (*DOG1*) regulates both seed dormancy and flowering time through microRNA pathways. Proc. Natl. Acad. Sci. USA 113(15): E2199–E2206. 10.1073/pnas.1600558113

Jablonka, E., Lamb, M.J. (2020) Inheritance Systems and the Extended Evolutionary Synthesis. Cambridge University Press, Cambridge, UK.

Jacques, M.A., et al. (2021) CODEX, a neural network approach to explore signaling dynamics landscapes. Molec. Syst. Biol. 17(4): e10026. 10.15252/msb.202010026

Janssens, P.M., Van Haastert, P.J. (1987) Molecular basis of transmembrane signal transduction in *Dictyostelium discoideum*. Microbiol. Rev. 51(4): 396–418. 10.1128/mmbr.51.4.396-418.1987

Jean-Baptiste, K., et al. (2019) Dynamics of gene expression in single root cells of *Arabidopsis thaliana*. Plant Cell 31(5): 993–1011. 10.1105/tpc.18.00785

Jiang-Xie, L.-F., et al. (2024) Neuronal dynamics direct cerebrospinal fluid perfusion and brain clearance. Nature 627(8002): 157–164. 10.1038/s41586-024-07108-6

Jing, H., et al. (2022) Regulation of AUXIN RESPONSE FACTOR condensation and nucleo-cytoplasmic partitioning. Nature Comm. 13(1): 4015. 10.1038/s41467-022-31628-2

Johansen, N., et al. (2023) Interindividual variation in human cortical cell type abundance and expression. Science 382(6667): eadf2359. 10.1126/science.adf2359

Johnston, I.G., Bassel, G.W. (2018) Identification of a bet-hedging network motif generating noise in hormone concentrations and germination propensity in Arabidopsis. J. Royal Soc. Interface 20180042(1): 1–12. 10.1098/rsif.2018.0042

Jores, T., et al. (2024) Plant enhancers exhibit both cooperative and additive interactions among their functional elements. Plant Cell 36(7): 2570–2586. 10.1093/plcell/koae088

Kagawa, T., Wada, M. (2004) Velocity of chloroplast avoidance movement is fluence rate dependent. Photochem. Photobiol. Sci. 3(6): 592–595. 10.1039/b316285k

Kaminski, A., et al. (2023) Resident regulatory T cells reflect the immune history of individual lymph nodes. Science Immun. 8(89): eadj5789. 10.1126/sciimmunol.adj5789

Kanatani, S., et al. (2024) Whole-brain spatial transcriptional analysis at cellular resolution. Science 386(6724): 907–915. 10.1126/science.adn9947

Kapanidis, A.N., et al. (2026) From sequence to function: Bridging single-molecule kinetics and molecular diversity. Science 391(6784): 458–465: 10.1126/science.adv4503

Karaiskos, N., et al. (2017) The *Drosophila* embryo at single-cell transcriptome resolution. Science 358(6360): 194–199. 10.1126/science.aan3235

Karmakar, R. (2021) State of the art of bacterial chemotaxis. J. Basic Microbiol. 61(5): 366–379. 10.1002/jobm.202000661

Kauffman, S.A. (1993) The Origins of Order: Self-Organization and Selection in Evolution. Oxford University Press, New York.

Kauffman, S.A. (2019) A World Beyond Physics: The Emergence and Evolution of Life. Oxford University Press, New York.

Kauffman, S.A., Roli, A. (2023) Beyond the Newtonian paradigm: A statistical mechanics of emergence. *In* PA Corning, SA Kauffman, D Noble, JA Shapiro, RI Vane-Wright, A Pross, eds, Evolution "On Purpose." Teleonomy in Living Systems. MIT Press, Cambridge, MA.

Kaur, U., et al. (2025) Autoinhibition imposed by a large conformational switch of INO80 regulates nucleosome positioning. Science 389(6757): eadr3831. 10.1126/science.adr3831

Kean-Galeno, T., et al. (2024) The shoot apical meristem: An evolutionary molding of higher plants. Int. J. Mol. Sci. 25(3): 1519. 10.3390/ijms25031519

Keung, A.J., Khalil, A.S. (2016) A unifying model of epigenetic regulation. Science 351(6274): 661–662. 10.1126/science.aaf1647

Khersonsky, O., Tawfik, D.S. (2010) Enzyme promiscuity: A mechanistic and evolutionary perspective. Annu. Rev. Biochem. 79(1): 471–505. 10.1146/annurev-biochem-030409-143718

Kilgore, H.R., et al. (2025) Protein codes promote selective subcellular compartmentalization. Science 387(6738): 1095–1101. 10.1126/science.adq2634

Kim, S.C., et al. (2021) The emergence of RNA from the heterogeneous products of prebiotic nucleotide synthesis. J. Am. Chem. Soc. 143(9): 3267–3279. 10.1021/jacs.0c12955

King, M.R., et al. (2024) Emergent microenvironments of nucleoli. Nucleus 15(1): 2319957. 10.1080/19491034.2024.2319957

Kinney, B., et al. (2023) A circadian-like gene network programs the timing and dosage of heterochronic miRNA transcription during *C. elegans* development. Devel. Cell 58(22): 2563–2579. 10.1016/j.devcel.2023.08.006

Kirschenbaum, D., et al. (2023) Time-resolved single-cell transcriptomics defines immune trajectories in glioblastoma. Cell 187(1): 149–165. 10.1016/j.cell.2023.11.032

Klingenberg, C.P. (2003) A developmental perspective on developmental instability: theory, models, and mechanisms. *In* M Polak, ed, Developmental Instability. Causes and Consequences. Oxford University Press, New York, pp 14–34.

Klingenberg, C.P. (2019) Phenotypic plasticity, developmental instability, and robustness: The concepts and how they are connected. Front. Ecol. Evol. 7(56): 1–15. 10.3389/fevo.2019.00056

Klingenberg, C.P., Nijhout, H.F. (1999) Genetics of fluctuating asymmetry: A developmental model of developmental instability. Evolution 53(2): 358–375. 10.1111/j.1558-5646.1999.tb03772.x

Klosin, A., et al. (2020) Phase separation provides a mechanism to reduce noise in cells. Science 367(6476): 464–468. 10.1126/science.aav6691

Komander, D., Rape, M. (2012) The ubiquitin code. Annu. Rev. Biochem. 81(1): 203–229. 10.1146/annurev-biochem-060310-170328

Kong, S.-G., et al. (2024) CHLOROPLAST UNUSUAL POSITIONING 1 is a plant-specific actin polymerization factor regulating chloroplast movement. Plant Cell 36(4): 1159–1181. 10.1093/plcell/koad320

Korobkova, E., et al. (2004) From molecular noise to behavioural variability in a single bacterium. Nature 428(6982): 574–578. 10.1038/nature02404

Koshland, D.E. (1980) Bacterial chemotaxis in relation to neurobiology. Annu. Rev. Neurosci. 3(1): 43–75. 10.1146/annurev.ne.03.030180.000355

Kou, K., et al. (2014) Islet number rather than islet size is a major determinant of β- and α-cell mass in humans. J. Clin. Endocrin. Metab. 99(5): 1733–1740. 10.1210/jc.2013-3731

Kramer, B.A., et al. (2022) Multimodal perception links cellular state to decision-making in single cells. Science 377(6606): 642–648. doi:10.1126/science.abf4062

Kreft, J.-U., et al. (2013) Mighty small: observing and modeling individual microbes becomes big science. Proc. Natl. Acad. Sci. USA 110(45): 18027–18028. 10.1073/pnas.1317472110

Krzyszton, M., et al. (2022) Single seeds exhibit transcriptional heterogeneity during secondary dormancy induction. Plant Physiol. 190(1): 211–225. 10.1093/plphys/kiac265

Krzyszton, M., et al. (2024) Dormancy heterogeneity among *Arabidopsis thaliana* seeds is linked to individual seed size. Plant Comm. 5(2): 100732. 10.1016/j.xplc.2023.100732

Ku, C., et al. (2015) Endosymbiotic origin and differential loss of eukaryotic genes. Nature 524(7566): 427–432. 10.1038/nature14963

Kuhn, T.S. (1962) The Structure of Scientific Revolutions. University of Chicago, Chicago.

Kull, K. (2022) Choices by organisms: On the role of freedom in behaviour and evolution. Biol. J. Linnean Soc. 139(4): 555–562. 10.1093/biolinnean/blac077

Kvenvolden, K., et al. (1970) Evidence for extraterrestrial amino-acids and hydrocarbons in the Murchison meteorite. Nature 228(5275): 923–926. 10.1038/228923a0

Laisk, A.G.U. (1983) Calculation of leaf photosynthetic parameters considering the statistical distribution of stomatal apertures. J. Exp. Bot. 34(12): 1627–1635. 10.1093/jxb/34.12.1627

Laisk, A.G.U., et al. (1980) Statistical distribution of stomatal apertures of *Vicia faba* and *Hordeum vulgare* and the Spannungsphase of stomatal opening. J. Exp. Bot. 31(1): 49–58. 10.1093/jxb/31.1.49

Lajus, D.L., et al. (2003) Developmental instability and the stochastic component of total phenotypic variance. *In* M Polak, ed, Developmental Instability: Causes and Consequences. Oxford University Press, New York, pp 343–363.

Lamon, S., et al. (2014) The STARS signaling pathway: a key regulator of skeletal muscle function. Pflügers Archiv – Euro. J. Physiol. 466(9): 1659–1671. 10.1007/s00424-014-1475-5

Landier, A., Thesmar, D. (2025) The Price of Our Values: The Economic Limits of Moral Life. University of Chicago Press, Chicago.

Lappalainen, J.K., et al. (2024) Connectome-constrained networks predict neural activity across the fly visual system. Nature 634(1): 1132–1140. 10.1038/s41586-024-07939-3

Larsson, A.J.M., et al. (2019) Genomic encoding of transcriptional burst kinetics. Nature 565(7738): 251–254. 10.1038/s41586-018-0836-1

Lauzon, D., Vallée-Bélisle, A. (2023) Programing chemical communication: Allostery vs multivalent mechanism. J. Am. Chem. Soc. 145(34): 18846–18854. 10.1021/jacs.3c04045
Lavaud, S., et al. (2024) Two inhibitory neuronal classes govern acquisition and recall of spinal sensorimotor adaptation. Science 384(6692): 194–202. 10.1126/science.adf6801
Leslie, M. (2024) Communication breakdown. Science 384(839–841.
Leveau, J.H.J., et al. (2018) Editorial: The individual microbe: single-cell analysis and agent-based modelling. Front. Microbiol. 9(2825): 1–3. 10.3389/fmicb.2018.02825
Lewis, S., et al. (2025) Scalable emulation of protein equilibrium ensembles with generative deep learning. Science 389(6761): eadv9817. 10.1126/science.adv9817
Li, M., Larsson, L. (2010) Force-generating capacity of human myosin isoforms extracted from single muscle fibre segments. J. Physiol. 588(24): 5105–5114. 10.1113/jphysiol.2010.199067
Liebherr, R.B., et al. (2014) A single molecule perspective on the functional diversity of *in vitro* evolved β-glucuronidase. J. Am. Chem. Soc. 136(16): 5949–5955. 10.1021/ja412379p
Liew, L.C., et al. (2024) Establishment of single-cell transcriptional states during seed germination. Nature Plants 10(9): 1418–1434. 10.1038/s41477-024-01771-3
Liu, A., et al. (2023) Encoding of environmental illumination by primate melanopsin neurons. Science 379(6630): 376–381. 10.1126/science.ade2024
Liu, L., et al. (2024) Enigmatic role of auxin response factors in plant growth and stress tolerance. Front. Plant Sci. 15(1398818): 1–24. 10.3389/fpls.2024.1398818
Liu, S., et al. (2020) Hydrothermal sensitivities of seed populations underlie fluctuations of dormancy states in an annual plant community. Ecology 101(3): e02958. 10.1002/ecy.2958
Liu, Z., et al. (2020) Global dynamic molecular profiling of stomatal lineage cell development by single-cell RNA sequencing. Mol. Plant 13(8): 1178–1193. 10.1016/j.molp.2020.06.010
Llamosi, A., et al. (2016) What population reveals about individual cell identity: Single-cell parameter estimation of models of gene expression in yeast. PLOS Computa. Biol. 12(2): e1004706. 10.1371/journal.pcbi.1004706
Lopez, B. (2022) Embrace Fearlessly the Burning World. Random House, New York.
Loveland, J.L., et al. (2025) A single gene orchestrates androgen variation underlying male mating morphs in ruffs. Science 387(6732): 406–412. 10.1126/science.adp5936
Lui, J.H., et al. (2011) Development and evolution of the human neocortex. Cell 146(1): 18–36. 10.1016/j.cell.2011.06.030
Luo, G., et al. (2016) Rapid oxygenation of Earth's atmosphere 2.33 billion years ago. Science Adv. 2(5): e1600134. 10.1126/sciadv.1600134
Luo, S., Zhou, X. (2025) Post-transcriptional regulation of behavior plasticity in social insects. Curr. Opin. Insect Sci. 68(1): 101329. 10.1016/j.cois.2024.101329
Lyon, P. (2015) The cognitive cell: Bacterial behavior reconsidered. Front. Microbiol. 6(264): 1–18. 10.3389/fmicb.2015.00264
Lyon, P., et al. (2021) Reframing cognition: getting down to biological basics. Phil. Trans. Royal Soc. B. Biol. Sci. 376(1820): 20190750. 10.1098/rstb.2019.0750
Ma, F., et al. (2021) Single-cell RNA sequencing of batch *Chlamydomonas* cultures reveals heterogeneity in their diurnal cycle phase. Plant Cell 33(4): 1042–1057. 10.1093/plcell/koab025
Ma, F., Zheng, C. (2023) Transcriptome age of individual cell types in *Caenorhabditis elegans*. Proc. Natl. Acad. Sci. USA 120(9): e2216351120. 10.1073/pnas.2216351120
Ma, R., et al. (2024) Principled and interpretable alignability testing and integration of single-cell data. Proc. Natl. Acad. Sci. USA 121(10): e2313719121. 10.1073/pnas.2313719121
Malavert, C., et al. (2021a) Light sensitivity changes during dormancy induction in *Polygonum aviculare* L. seeds: development of a predictive model of annual changes in seed-bank light sensitivity in relation to soil temperature. Weed Res. 61(2): 115–125. 10.1111/wre.12463

Malavert, C., et al. (2021b) The role of seed water content for the perception of temperature signals that drive dormancy changes in *Polygonum aviculare* buried seeds. Funct. Plant Biol. 48(1): 28–39. 10.1071/FP20011

Marand, A.P., et al. (2023) cis-Regulatory elements in plant development, adaptation, and evolution. Annu. Rev. Plant Biol. 74(1): 111–137. 10.1146/annurev-arplant-070122-030236

Marand, A.P., et al. (2025) The genetic architecture of cell type–specific cis regulation in maize. Science 388(6744): eads6601. 10.1126/science.ads6601

Martin, P.R. (2023) Neurons share an intense load. Science 379(6630): 335–336. 10.1126/science.adf9350

Martin-Arevalillo, R., Vernoux, T. (2023) Decoding the auxin matrix: auxin biology through the eye of the computer. Annu. Rev. Plant Biol. 74(1): 387–413. 10.1146/annurev-arplant-102720-033523

Mato, J.M., et al. (1975) Signal input for a chemotactic response in the cellular slime mold *Dictyostelium discoideum*. Proc. Natl. Acad. Sci. USA 72(12): 4991–4993. 10.1073/pnas.72.12.4991

McDonald, J.E., et al. (2020) Application of ecological and evolutionary theory to microbiome community dynamics across systems. Proc. Royal Soc. B. Biol. Sci. 287(1941): 20202886. 10.1098/rspb.2020.2886

McFadden, J., Al-Khalili, J. (2014) Life on the Edge. The Coming Age of Quantum Biology. Broadway Books, New York.

Merchant, S.S., et al. (2007) The *Chlamydomonas* genome reveals the evolution of key animal and plant functions. Science 318(5848): 245–250. 10.1126/science.1143609

Mesibov, R., et al. (1973) The range of attractant concentrations for bacterial chemotaxis and the threshold and size of response over this range. J. Gen. Physiol. 62(2): 203–223. 10.1085/jgp.62.2.203

Metcalf, C.J.E., Koskella, B. (2019) Protective microbiomes can limit the evolution of host pathogen defense. Evol. Lett. 3(5): 534–543. 10.1002/evl3.140

Metcalf, C.J.E., et al. (2025) Convergence and divergence of individual immune responses over the life course. Science 389(6760): 604–609. 10.1126/science.ady9543

Meyer, S.E., Allen, P.S. (2009) Predicting seed dormancy loss and germination timing for *Bromus tectorum* in a semi-arid environment using hydrothermal time models. Seed Sci. Res. 19(04): 225–239. 10.1017/S0960258509990122

Michelon, T.B., et al. (2025) Soybean single-seed respiration evaluation through spectral imaging. Seed Sci. Res. 35(3): 169–180. 10.1017/S0960258525100044

Mitchell, K.J. (2023) Free Agents. How Evolution Gave Us Free Will. Princeton University Press, Princeton, NJ.

Moltchanova, E., et al. (2020) Comparison of three different statistical approaches (non-linear least-squares regression, survival analysis and Bayesian inference) in their usefulness for estimating hydrothermal time models of seed germination. Seed Sci. Res. 30(1): 64–72. 10.1017/S0960258520000082

Monahan, K., Lomvardas, S. (2015) Monoallelic expression of olfactory receptors. Annu. Rev. Cell Dev. Biol. 31(1): 721–740. 10.1146/annurev-cellbio-100814-125308

Morris, J.P., et al. (2024) Integrating the study of polyploidy across organisms, tissues, and disease. Annu. Rev. Genet. 58: 297–318. 10.1146/annurevgenet-111523-102124

Mott, K.A., Buckley, T.N. (2000) Patchy stomatal conductance: emergent collective behaviour of stomata. Trends Plant Sci. 5(6): 258–262. 10.1016/S1360-1385(00)01648-4

Moussu, S., et al. (2023) Plant cell wall patterning and expansion mediated by protein-peptide-polysaccharide interaction. Science 382(6671): 719–725. 10.1126/science.adi4720

Mukherjee, S. (2022) The Song of the Cell: An Exploration of Medicine and the New Human. Simon and Schuster, New York.

Mukherji, S., et al. (2011) MicroRNAs can generate thresholds in target gene expression. Nature Genet. 43(9): 854–859. 10.1038/ng.905

Muljadi, M., et al. (2022) Understanding the cell's response to chemical signals: Utilisation of microfluidic technology in studies of cellular and *Dictyostelium discoideum* chemotaxis. Micromachines 13(10): 1737. 10.3390/mi13101737

Mund, A., et al. (2022) Deep Visual Proteomics defines single-cell identity and heterogeneity. Nat. Biotech. 40(1): 1231–1240. 10.1038/s41587-022-01302-5

Myers, N. (2015) Rendering Life Molecular. Duke University Press, Durham, NC.

Nagano, S. (2000) Modeling the model organism *Dictyostelium discoideum*. Devel. Growth Differentia. 42(6): 541–550. 10.1046/j.1440-169x.2000.00547.x

Nakamura, A., et al. (2021) Shedding light on developmental ERK signaling with genetically encoded biosensors. Development 148(dev199767): 1–11. 10.1242/dev.199767

Nakashima, K., et al. (2009) Three Arabidopsis SnRK2 protein kinases, SRK2D/SnRK2.2, SRK2E/SnRK2.6/OST1 and SRK2I/SnRK2.3, involved in ABA signaling are essential for the control of seed development and dormancy. Plant Cell Physiol. 50(7): 1345–1363. 10.1093/pcp/pcp083

Nakayama, H., et al. (2022) Molecular mechanisms underlying leaf development, morphological diversification, and beyond. Plant Cell 34(7): 2534–2548. 10.1093/plcell/koac118

Nalbant, P., et al. (2023) Direct investigation of cell contraction signal networks by light-based perturbation methods. Pflügers Archiv – Euro. J. Physiol. 475(12): 1439–1452. 10.1007/s00424-023-02864-2

Nelson, D.L., et al. (2008) Lehninger Principles of Biochemistry. Macmillan, New York.

Nemati, I., et al. (2022) *DELAY OF GERMINATION 1* (*DOG1*) regulates dormancy in dimorphic seeds of *Xanthium strumarium*. Funct. Plant Biol. 49(8): 742–758. 10.1071/FP21315

Neubig, R.R., et al. (2003) International Union of Pharmacology Committee on Receptor Nomenclature and Drug Classification. XXXVIII. Update on terms and symbols in quantitative pharmacology. Pharmacol. Rev. 55(1): 597–606. 10.1124/pr.55.4.4

Nguyen, E., et al. (2024) Sequence modeling and design from molecular to genome scale with Evo. Science 386(6723): eado9336. 10.1126/science.ado9336

Ni, B.R., Bradford, K.J. (1992) Quantitative models characterizing seed germination responses to abscisic acid and osmoticum. Plant Physiol. 98(3): 1057–1068. 10.1104/pp.98.3.1057

Ni, B.R., Bradford, K.J. (1993) Germination and dormancy of abscisic acid-deficient and gibberellin-deficient mutant tomato seeds. Sensitivity of germination to abscisic acid, gibberellin, and water potential. Plant Physiol. 101(2): 607–617. 10.1104/pp.101.2.607

Nickoloff, J.A., et al. (2021) The safe path at the fork: Ensuring replication-associated DNA double-strand breaks are repaired by homologous recombination. Front. Genet. 12(1): 1–9. 10.3389/fgene.2021.748033

Nijhout, H.F. (2002) The nature of robustness in development. Bioessays 24(6): 553–563. 10.1002/bies.10093

Nijhout, H.F., Davidowitz, G. (2003) Developmental perspectives on phenotypic variation, canalization, and fluctuating asymmetry. *In* M Polak, ed, Developmental Instability. Causes and Consequences. Oxford University Press, New York, pp 3–13.

Nijhout, H.F., et al. (2014) Escape from homeostasis. Math. Biosci. 257(1): 104–110. 10.1016/j.mbs.2014.08.015

Noble, D.W.A., et al. (2019) Plastic responses to novel environments are biased towards phenotype dimensions with high additive genetic variation. Proc. Natl. Acad. Sci. USA 116(27): 13452–13461. 10.1073/pnas.1821066116

Noble, R., Noble, D. (2018) Harnessing stochasticity: How do organisms make choices? Chaos: An Interdisc. J. Nonlinear Sci. 28(10): 1–7. 10.1063/1.5039668

Nonogaki, H. (2020) A repressor complex silencing ABA signaling in seeds? J. Exp. Bot. 71(10): 2847–2853. 10.1093/jxb/eraa062

Nonogaki, H., et al. (2000) A germination-specific endo-β-mannanase gene is expressed in the micropylar endosperm cap of tomato seeds. Plant Physiol. 123(4): 1235–1245. 10.1104/pp.123.4.1235

Nonogaki, H., et al. (2007) Mechanisms and genes involved in germination *sensu stricto*. *In* KJ Bradford, H Nonogaki, eds, Seed Development, Dormancy and Germination. Blackwell Publishing, Oxford, pp 264–304.

Norman-Haignere, S.V., et al. (2022) A neural population selective for song in human auditory cortex. Curr. Biol. 32(7): 1470–1484.e1412. 10.1016/j.cub.2022.01.069

Nowak, M.A., et al. (1997) Evolution of genetic redundancy. Nature 388(6638): 167–171. 10.1038/40618

Nusinovich, Y. (2024) Decoding the brain. Science 384(6698): 858–859. 10.1126/science.adp9365

Nyman, J., et al. (2023) Spatially aware deep learning reveals tumor heterogeneity patterns that encode distinct kidney cancer states. Cell Rep. Med. 4(9): 101189. 10.1016/j.xcrm.2023.101189

Ohtsuka, D., et al. (2021) A sub-population of *Dictyostelium discoideum* cells shows extremely high sensitivity to cAMP for directional migration. Biochem. Biophys. Res. Comm. 554(1): 131–137. 10.1016/j.bbrc.2021.03.095

Onofri, A., et al. (2014) Experimental design and parameter estimation for threshold models in seed germination. Weed Res. 54(5): 425–435. 10.1111/wre.12095

Orgel, L.E. (2004) Prebiotic chemistry and the origin of the RNA world. Crit. Rev. Biochem. Mol. Biol. 39(2): 99–123. 10.1080/10409230490460765

Ouonkap, S.V.Y., et al. (2024) Enhanced pollen tube performance at high temperature contributes to thermotolerant fruit and seed production in tomato. Curr. Biol. 34(22): 5319–5333. 10.1016/j.cub.2024.10.025

Papadopoulos, D.K., et al. (2019) Control of Hox transcription factor concentration and cell-to-cell variability by an auto-regulatory switch. Development 146(12): 1–15. 10.1242/dev.168179

Papastavrou, N., et al. (2024) RNA-catalyzed evolution of catalytic RNA. Proc. Natl. Acad. Sci. USA 121(11): e2321592121. 10.1073/pnas.2321592121

Parent, C.A., Devreotes, P.N. (1999) A cell's sense of direction. Science 284(5415): 765–770. 10.1126/science.284.5415.765

Parthasarathy, R. (2022) So Simple a Beginning: How Four Physical Principles Shape Our Living World. Princeton University Press, Princeton, NJ.

Pearce, J.D., et al. (2025) A cross-species generative cell atlas across 1.5 billion years of evolution: The TranscriptFormer single-cell model. bioRxiv: 2025.2004.2025.650731. 10.1101/2025.04.25.650731

Pelletier, J.M., et al. (2025) Dissecting the cellular architecture and genetic circuitry of the soybean seed. Proc. Natl. Acad. Sci. USA 122(1): e2416987121. 10.1073/pnas.2416987121

Pennisi, E. (2018) The power of many. Science 360(6396): 1388–1391. 10.1126/science.360.6396.1388

Pezzotta, A., Briscoe, J. (2023) Optimal control of gene regulatory networks for morphogen-driven tissue patterning. Cell Systems 14(11): 940–952. 10.1016/j.cels.2023.10.004

Pinker, S. (2021) Rationality: What It Is, Why It Seems Scarce, Why It Matters. Viking, New York.

Piot, L., et al. (2023) GluD1 binds GABA and controls inhibitory plasticity. Science 382(6677): 1389–1394. 10.1126/science.adf3406

Pipeleers, D. (1987) The biosociology of pancreatic B cells. Diabetologia 30(5): 277–291. 10.1007/BF00299019

Pipeleers, D., et al. (2017) Heterogeneity in the beta-cell population: A guided search into its significance in pancreas and in implants. Curr. Diabetes Rep. 17(10): 86. 10.1007/s11892-017-0925-9

Pipeleers, D.G. (1992) Heterogeneity in pancreatic β-cell population. Diabetes 41(7): 777–781. 10.2337/diab.41.7.777

Plavskin, Y., et al. (2016) Ancient trans-acting siRNAs confer robustness and sensitivity onto the auxin response. Devel. Cell 36(3): 276–289. 10.1016/j.devcel.2016.01.010

Polak, M., ed (2003) Developmental Instability. Causes and Consequences. Oxford University Press, New York.

Posadas, D.M., Carthew, R.W. (2014) MicroRNAs and their roles in developmental canalization. Curr. Opin. Genet. Dev. 27(1): 1–6. 10.1016/j.gde.2014.03.005

Pountain, A.W., Yanai, I. (2025) Dissecting microbial communities with single-cell transcriptome analysis. Science 389(6764): eadp6252. 10.1126/science.adp6252

Pourmorady, A.D., et al. (2024) RNA-mediated symmetry breaking enables singular olfactory receptor choice. Nature 625(7993): 181–188. 10.1038/s41586-023-06845-4

Powers, S.K., et al. (2019) Nucleo-cytoplasmic partitioning of ARF proteins controls auxin responses in *Arabidopsis thaliana.* Mol. Cell 76(1): 177–190. 10.1016/j.molcel.2019.06.044

Prest, B.C., et al. (2024) Equity weighting increases the social cost of carbon. Science 385(6710): 715–717. 10.1126/science.adn1488

Purvis, J.E., Lahav, G. (2013) Encoding and decoding cellular information through signaling dynamics. Cell 152(5): 945–956. 10.1016/j.cell.2013.02.005

Qiao, Z. (2025) AI to rewire life's interactome: Structural foundation models help to elucidate and reprogram molecular biology. Science 389(6757): 244–245. 10.1126/science.adx7802

Radersma, R., et al. (2020) Plasticity leaves a phenotypic signature during local adaptation. Evol. Lett. 4(4): 360–370. 10.1002/evl3.185

Railsback, S.F., Grimm, V. (2019) Agent-Based and Individual-Based Modeling: A Practical Introduction, Second Edition. Princeton University Press, Princeton, NJ.

Rajan, D., et al. (2023) Single-cell analysis of habituation in *Stentor coeruleus.* Curr. Biol. 33(2): 241–251. 10.1016/j.cub.2022.11.010

Rajapakshe, R.P.V.G.S.W., et al. (2024) Same, same, but different: dissimilarities in the hydrothermal germination performance of range-restricted endemics emerge despite microclimatic similarities. Conserv. Physiol. 12(1): coae009. 10.1093/conphys/coae009

Ram, A., et al. (2023a) A guide to ERK dynamics, part 1: mechanisms and models. Biochem. J. 480(23): 1887–1907. 10.1042/bcj20230276

Ram, A., et al. (2023b) A guide to ERK dynamics, part 2: downstream decoding. Biochem. J. 480(23): 1909–1928. 10.1042/bcj20230277

Ramat, A., Simonelig, M. (2022) Activating translation with phase separation. Science 377(6607): 712–713. 10.1126/science.add6323

Raser, J.M., O'Shea, E.K. (2005) Noise in gene expression: Origins, consequences, and control. Science 309(5743): 2010–2013. 10.1126/science.1105891

Reed, R.C., et al. (2022) Seed germination and vigor: ensuring crop sustainability in a changing climate. Heredity 128(450–459. 10.1038/s41437-022-00497-2

Remus-Emsermann, M.N.P., Leveau, J.H.J. (2010) Linking environmental heterogeneity and reproductive success at single-cell resolution. ISME Journal 4(2): 215–222. 10.1038/ismej.2009.110

Richards, T.A., et al. (2019) Single cell ecology. Phil. Trans. Royal Soc. B. Biol. Sci. 374(1786): 20190076. 10.1098/rstb.2019.0076

Roberts, E.H. (1960) The viability of cereal seed in relation to temperature and moisture. Ann. Bot. 24(1): 12–31. 10.1093/oxfordjournals.aob.a083684

Roberts, E.H. (1961) Dormancy of rice seed. I. The distribution of dormancy periods. J. Exp. Bot. 12(2): 319–329. 10.1093/jxb/12.2.319

Roberts, E.H. (1973) Predicting the storage life of seeds. Seed Sci. Technol. 1(3): 499–514.

Roberts, E.H., Ellis, R.H. (1989) Water and seed survival. Ann. Bot. 63(1): 39–52. 10.1093/oxfordjournals.aob.a087727

Rodbard, D. (1973) Theory of hormone-receptor interaction. III. The endocrine target cell as a quantal response unit: a general control mechanism. Adv. Exp. Med. Biol. 36 (1): 342–364.

Romero-Perez, P.S., et al. (2023) When phased without water: Biophysics of cellular desiccation, from biomolecules to condensates. Chem. Rev. 123(44): 9010–9035. 10.1021/acs.chemrev.2c00659

Rood, J.E., et al. (2024) The Human Cell Atlas from a cell census to a unified foundation model. Nature 637(1): 1065–1071. 10.1038/s41586-024-08338-4

Rowe, J., et al. (2023) Next-generation ABACUS biosensors reveal cellular ABA dynamics driving root growth at low aerial humidity. Nature Plants 9(7): 1103–1115. 10.1038/s41477-023-01447-4

Rubio de Casas, R., et al. (2015) Gene-flow through space and time: dispersal, dormancy and adaptation to changing environments. Evol. Ecol. 29(6): 813–831. 10.1007/s10682-015-9791-6

Russ, D.E., et al. (2021) A harmonized atlas of mouse spinal cord cell types and their spatial organization. Nature Comm. 12(1): 5722. 10.1038/s41467-021-25125-1

Salomé, P.A., Merchant, S.S. (2021) Co-expression networks in *Chlamydomonas* reveal significant rhythmicity in batch cultures and empower gene function discovery. Plant Cell 33(4): 1058–1082. 10.1093/plcell/koab042

Salome, P.A., et al. (2008) Circadian timekeeping during early Arabidopsis development. Plant Physiol. 147(3): 1110–1125. 10.1104/pp.108.117622

Sands, L.P., et al. (2026) Octopamine and tyramine dynamics predict learning rate phenotypes during associative conditioning in honey bees. Science Advances 12(7): eaea8433. 10.1126/sciadv.aea8433

Santoni, G., et al. (2024) Chromatin plasticity predetermines neuronal eligibility for memory trace formation. Science 385(6707): eadg9982. 10.1126/science.adg9982

Sapolsky, R.M. (2023) Determined. A Science of Life without Free Will. Penguin Press, New York.

Scacchi, E., et al. (2024) A diffusible small-RNA-based Turing system dynamically coordinates organ polarity. Nature Plants 10(3): 412–422. 10.1038/s41477-024-01634-x

Schlegel, P., et al. (2024) Whole-brain annotation and multi-connectome cell typing of *Drosophila*. Nature 634(8032): 139–152. 10.1038/s41586-024-07686-5

Schmiedel, J.M., et al. (2015) MicroRNA control of protein expression noise. Science 348(6230): 128–132. 10.1126/science.aaa1738

Schneider, T., et al. (2023) Specifying conformational heterogeneity of multi-domain proteins at atomic resolution. Structure 31(10): 1259–1274. 10.1016/j.str.2023.07.008

Schott, M., et al. (2024) Open-ST: High-resolution spatial transcriptomics in 3D. Cell 187(15): 3953–3972. 10.1016/j.cell.2024.05.055

Schramma, N., et al. (2023) Chloroplasts in plant cells show active glassy behavior under low-light conditions. Proc. Natl. Acad. Sci. USA 120(3): e2216497120. 10.1073/pnas.2216497120

Sekar, V., et al. (2023) Detection of transcriptome-wide microRNA–target interactions in single cells with agoTRIBE. Nat. Biotech. 42(1): 1296–1302. 10.1038/s41587-023-01951-0

Seller, C.A., Schroeder, J.I. (2023) Distinct guard cell-specific remodeling of chromatin accessibility during abscisic acid- and CO_2-dependent stomatal regulation. Proc. Natl. Acad. Sci. USA 120(52): e2310670120. 10.1073/pnas.2310670120

Sente, A. (2024) Gatekeepers of the brain: Identifying hidden mechanisms of type A GABA receptor signaling and assembly. Science 386(6723): 738–739. 10.1126/science.adt8990

Serrano, K., et al. (2024) Unraveling plant-microbe symbioses using single-cell and spatial transcriptomics. Trends Plant Sci. 29(12): 1356–1367. 10.1016/j.tplants.2024.06.008

Shalek, A.K., et al. (2013) Single-cell transcriptomics reveals bimodality in expression and splicing in immune cells. Nature 498(7453): 236–240. 10.1038/nature12172

Shi, X., Ling, H.-Q. (2018) Current advances in genome sequencing of common wheat and its ancestral species. Crop J. 6(1): 15–21. 10.1016/j.cj.2017.11.001

Siciliano, V., et al. (2013) miRNAs confer phenotypic robustness to gene networks by suppressing biological noise. Nature Comm. 4(1): 2364. 10.1038/ncomms3364

Siletti, K., et al. (2023) Transcriptomic diversity of cell types across the adult human brain. Science 382(6667): eadd7046. 10.1126/science.add7046

Sinha, S., et al. (2024) PERCEPTION predicts patient response and resistance to treatment using single-cell transcriptomics of their tumors. Nature Cancer 5(1): 938–952. 10.1038/s43018-024-00756-7

Sirichandra, C., et al. (2009) The guard cell as a single-cell model towards understanding drought tolerance and abscisic acid action. J. Exp. Bot. 60(5): 1439–1463. 10.1093/jxb/ern340

Skopelitis, D.S., et al. (2012) Plant small RNAs as morphogens. Curr. Opin. Cell Biol. 24(2): 217–224. 10.1016/j.ceb.2011.12.006

Skopelitis, D.S., et al. (2017) Boundary formation through a direct threshold-based readout of mobile small RNA gradients. Devel. Cell 43(3): 265–273. 10.1016/j.devcel.2017.10.003

Sliwinska, E., et al. (2009) Germination of *Arabidopsis thaliana* seeds is not completed as a result of elongation of the radicle but of the adjacent transition zone and lower hypocotyl. J. Exp. Bot. 60(12): 3587–3594. 10.1093/jxb/erp203

Solis-Miranda, J., et al. (2023) Stress-related biomolecular condensates in plants. Plant Cell 35(9): 3187–3204. 10.1093/plcell/koad127

Soltani, E., et al. (2015) Problems with using mean germination time to calculate rate of seed germination. Aust. J. Bot. 63(8): 631–635. 10.1071/BT15133

Sparta, B., et al. (2015) Receptor level mechanisms are required for epidermal growth factor (EGF)-stimulated extracellular signal-regulated kinase (ERK) activity pulses. J. Biol. Chem. 290(41): 24784–24792. 10.1074/jbc.M115.662247

Springthorpe, V., Penfield, S. (2015) Flowering time and seed dormancy control use external coincidence to generate life history strategy. eLife 4(1): e05557. 10.7554/eLife.05557

Spudich, J.L., Koshland, D.E. (1976) Non-genetic individuality: chance in the single cell. Nature 262(5568): 467–471. 10.1038/262467a0

Stadler, M.R., Eisen, M.B. (2017) Atlas...t, patterns from every cell. Science 358(6360): 172–173. 10.1126/science.aap8493

Stamp, M.A., Hadfield, J.D. (2020) The relative importance of plasticity versus genetic differentiation in explaining between population differences; a meta-analysis. Ecol. Lett. 23(10): 1432–1441. 10.1111/ele.13565

Steens, J.A., et al. (2024) Type III-B CRISPR-Cas cascade of proteolytic cleavages. Science 383(6682): 512–519. 10.1126/science.adk0378

Still, D.W., Bradford, K.J. (1997) Endo-β-mannanase activity from individual tomato endosperm caps and radicle tips in relation to germination rates. Plant Physiol. 113(1): 21–29. 10.1104/pp.113.1.21

Stinchcombe, J.R., Kirkpatrick, M. (2012) Genetics and evolution of function-valued traits: understanding environmentally responsive phenotypes. Trend. Ecol. Evol. 27(11): 637–647. 10.1016/j.tree.2012.07.002

Strenkert, D., et al. (2019) Multiomics resolution of molecular events during a day in the life of *Chlamydomonas.* Proc. Natl. Acad. Sci. USA 116(6): 2374–2383. 10.1073/pnas.1815238116

Su, J.-H., et al. (2020) Genome-scale imaging of the 3D organization and transcriptional activity of chromatin. Cell 182(6): 1641–1659. 10.1016/j.cell.2020.07.032

Suetsugu, N., Wada, M. (2020) Signalling mechanism of phototropin-mediated chloroplast movement in Arabidopsis. J. Plant Biochem. Biotech. 29(4): 580–589. 10.1007/s13562-020-00594-5

Szostak, J.W. (2017) The narrow road to the deep past: In search of the chemistry of the origin of life. Angewandte Chemie Internat. Ed. 56(37): 11037–11043. 10.1002/anie.201704048

Szostak, J.W., et al. (2001) Synthesizing life. Nature 409(6818): 387–390. 10.1038/35053176

Tameshige, T., et al. (2025) Mutual inhibition between EPFL2 and auxin extends the intervals of periodic leaf morphogenesis. Nature Communications 16(1): 9753. 10.1038/s41467-025-65792-y

Tan, B., et al. (2024) Design of a SARS-CoV-2 papain-like protease inhibitor with antiviral efficacy in a mouse model. Science 383(6690): 1434–1440. 10.1126/science.adm9724

Tanday, N., et al. (2024) Pancreatic islet cell plasticity: Pathogenic or therapeutically exploitable? Diabetes Obesity Metab. 26(1): 16–31. 10.1111/dom.15300

Taniguchi, Y., et al. (2010) Quantifying *E. coli* proteome and transcriptome with single-molecule sensitivity in single cells. Science 329(5991): 533–538. 10.1126/science.1188308

TechnologyNetworks (2024) Fresh evidence for the "RNA World" hypthesis. https://www.technologynetworks.com/genomics/news/fresh-evidence-for-the-rna-world-hypothesis-384456

ten Brink, H., et al. (2020) Optimal germination timing in unpredictable environments: the importance of dormancy for both among- and within-season variation. Ecol. Lett. 23(4): 620–630. 10.1111/ele.13461

Terashima, I., et al. (1988) Characterisation of non-uniform photosynthesis induced by abscisic acid in leaves having different mesophyll anatomies. Plant Cell Physiol. 29(3): 385–394. 10.1093/oxfordjournals.pcp.a077506

Thiebaut de Schotten, M., Forkel, S.J. (2022) The emergent properties of the connected brain. Science 378(6619): 505–510. 10.1126/science.abq2591

Tian, C., et al. (2020) Single-nucleus RNA-seq resolves spatiotemporal developmental trajectories in the tomato shoot apex. bioRxiv: 2020.2009.2020.305029. 10.1101/2020.09.20.305029

Topham, A.T., et al. (2017) Temperature variability is integrated by a spatially embedded decision-making center to break dormancy in *Arabidopsis* seeds. Proc. Natl. Acad. Sci. USA 114(25): 6629–6634. 10.1073/pnas.1704745114

Tranquillo, R., et al. (1988) A stochastic model for leukocyte random motility and chemotaxis based on receptor binding fluctuations. J. Cell Biol. 106(2): 303–309. 10.1083/jcb.106.2.303

Trewavas, A. (1999) The importance of individuality. *In* HR Lerner, ed, Plant Responses to Environmental Stresses. From Phytohormones to Genome Reorganization. Marcel Dekker, New York, pp 27–42.

Trewavas, A.J. (2012) Information, noise and communication: thresholds as controlling elements in development. *In* G Witzany, F Baluska, eds, Biocommunication of Plants. Signaling and Communication in Plants vol. 14. Springer-Verlag, Berlin, pp 11–35.

Trewavas, A.J. (2014) Plant Behaviour and Intelligence. Oxford University Press, Oxford, UK.

Trewavas, A.J., Cleland, R.E. (1983) Is plant development regulated by changes in the concentration of growth substances or by changes in the sensitivity to growth substances? Trends Biochem. Sci. 8(10): 354–357.

Trudgill, D.L., et al. (2005) Thermal time—concepts and utility. Ann. Appl. Biol. 146(1): 1–14. 10.1111/j.1744-7348.2005.04088.x

Tsimring, L.S. (2014) Noise in biology. Rep. Prog. Physics 77(2): 026601. 10.1088/0034-4885/77/2/026601

Turing, A.M. (1952) The chemical basis of morphogenesis. Phil. Trans. R. Soc. Lond. B. Biol. Sci. 237(1): 37–72.

Umezawa, T., et al. (2009) Type 2C protein phosphatases directly regulate abscisic acid-activated protein kinases in Arabidopsis. Proc. Natl. Acad. Sci. USA 106(41): 17588–17593. 10.1073/pnas.0907095106

Valdeolmillos, M., et al. (1996) *In vivo* synchronous membrane potential oscillations in mouse pancreatic beta-cells: lack of co-ordination between islets. J. Physiol. 493(1): 9–18. 10.1113/jphysiol.1996.sp021361

Van De Winkel, M., Pipeleers, D. (1983) Autofluorescence-activated cell sorting of pancreatic islet cells: Purification of insulin-containing B-cells according to glucose-induced changes in cellular redox state. Biochem. Biophys. Res. Comm. 114(2): 835–842. 10.1016/0006-291X(83)90857-4

Van Schravendijk, C.F., et al. (1992) Pancreatic beta cell heterogeneity in glucose-induced insulin secretion. J. Biol. Chem. 267(30): 21344–21348. 10.1016/S0021-9258(19)36615-3

Vane-Wright, R.I. (2023) Turning biology to life: some reflections. Biol. J. Linnean Soc. 139(4): 570–587. 10.1093/biolinnean/blac141

Venable, D.L. (2007) Bet hedging in a guild of desert annuals. Ecology 88(5): 1086–1190. 10.1890/06-1495

Venable, D.L., Lawlor, L. (1980) Delayed germination and dispersal in desert annuals: Escape in space and time. Oecologia 46(2): 272–282. 10.1007/BF00540137

Vidigal, D.S., et al. (2016) Altitudinal and climatic associations of seed dormancy and flowering traits evidence adaptation of annual life cycle timing in *Arabidopsis thaliana*. Plant Cell. Environ. 39(8): 1737–1748. 10.1111/pce.12734

Vilches, L.F., Dickinson, J.D. (2024) Start SPREADing the news: biosensors detect ripples of ERK signaling in airway epithelial cells. Am. J. Respir. Cell Mol. Biol. 72(5): 469–471. 10.1165/rcmb.2024-0547ED

Wada, M., Kong, S.-G. (2018) Actin-mediated movement of chloroplasts. J. Cell Sci. 131(2): jcs210310. 10.1242/jcs.210310

Waddington, C.H. (1942) Canalization of development and the inheritance of acquired characters. Nature 150(3811): 563–565. 10.1038/150563a0

Wagner, D.E., et al. (2018) Single-cell mapping of gene expression landscapes and lineage in the zebrafish embryo. Science 360(6392): 981–987. 10.1126/science.aar4362

Walde, M.G., et al. (2025) Rethinking the interpretation of spring phenological temperature sensitivity. npj Sci. Plants 1(3): 1–3. 10.1038/s44383-025-00004-6

Wang, C.J., et al. (2012) Diverse sensitivity thresholds in dynamic signaling responses by social amoebae. Science Signal. 5(213): 1–11. 10.1126/scisignal.2002449

Wang, X., et al. (2024) Noncanonical transcription initiation is primarily tissue specific and epigenetically tuned in paleopolyploid plants. Plant Cell 37(1): 1–15. 10.1093/plcell/koae288

Watahiki, M., Trewavas, A. (2019) Systems, variation, individuality and plant hormones. Prog. Biophys. Mol. Biol. 146(1): 3–22. 10.1016/j.pbiomolbio.2018.10.001

Waterworth, W.M., West, C.E. (2023) Genome damage accumulated in seed ageing leads to plant genome instability and growth inhibition. Biochem. J. 480(7): 461–470. 10.1042/bcj20230006

Waterworth, W.M., et al. (2016) DNA damage checkpoint kinase ATM regulates germination and maintains genome stability in seeds. Proc. Natl. Acad. Sci. USA 113(34): 9647–9652. 10.1073/pnas.1608829113

Waterworth, W.M., et al. (2019) Seeds and the art of genome maintenance. Front. Plant Sci. 10(706): 1–11. 10.3389/fpls.2019.00706

Watson, J.D., Crick, F.H.C. (1953) Molecular structure of nucleic acids: A structure for deoxyribose nucleic acid. Nature 171(4356): 737–738. 10.1038/171737a0

Watson, J.L., et al. (2023) Macromolecular condensation buffers intracellular water potential. Nature 623(7988): 842–852. 10.1038/s41586-023-06626-z

Weis, K., Hondele, M. (2022) The role of DEAD-Box ATPases in gene expression and the regulation of RNA–protein condensates. Annu. Rev. Biochem. 91(1): 197–219. 10.1146/annurev-biochem-032620-105429

Weng, J.-K., et al. (2012) The rise of chemodiversity in plants. Science 336(6089): 1667–1670. 10.1126/science.1217411

West-Eberhard, M.J. (2003) Foreword. *In* M Polak, ed, Developmental Instability. Causes and Consequences. Oxford University Press, New York.

Williams, R.J. (1956) Biochemical Individuality. The Basis for the Genetotrophic Concept. John Wiley & Sons, New York.

Williams, T.D. (2008) Individual variation in endocrine systems: moving beyond the 'tyranny of the Golden Mean.' Phil. Trans. Royal Soc. B. Biol. Sci. 363(1497): 1687–1698. 10.1098/rstb.2007.0003

Wolkis, D., et al. (2025) Seed longevity: analysing post-storage germination data in R to fit the viability equation. Seed Sci. Res. 35(1): 33–40. 10.1017/S0960258524000291

Wong, M.L., et al. (2023) On the roles of function and selection in evolving systems. Proc. Natl. Acad. Sci. USA 120(43): e2310223120. 10.1073/pnas.2310223120

Wood, D.C. (1969) Parametric studies of the response decrement produced by mechanical stimuli in the protozoan, *Stentor coeruleus*. J. Neurobiol. 1(3): 345–360. 10.1002/neu.480010309

Wright, W.J., et al. (2025) Distinct synaptic plasticity rules operate across dendritic compartments *in vivo* during learning. Science 388(6744): 322–328. 10.1126/science.ads4706

Wu, C.-I., et al. (2009) Evolution under canalization and the dual roles of microRNAs—A hypothesis. Genome Res. 19(1): 734–743. 10.1101/gr.084640.108

Wu, K., et al. (2025) Design of intrinsically disordered region binding proteins. Science 389(6757): eadr8063. 10.1126/science.adr8063

Xavier, J.C., et al. (2020) Autocatalytic chemical networks at the origin of metabolism. Proc. Royal Soc. B. Biol. Sci. 287(1922): 20192377. 10.1098/rspb.2019.2377

Xia, K., et al. (2022) The single-cell stereo-seq reveals region-specific cell subtypes and transcriptome profiling in Arabidopsis leaves. Developmental Cell 57(10): 1299–1310.e1294. 10.1016/j.devcel.2022.04.011

Xue, Z., et al. (2024) Natural variation in response to combined water and nitrogen deficiencies in Arabidopsis. Plant Cell 36(9): 3378–3398. 10.1093/plcell/koae173

Yamamoto, R., et al. (2024) Hyphae of the fungus *Aspergillus nidulans* demonstrate chemotropism to nutrients and pH. PLOS Biology 22(7): e3002726. 10.1371/journal.pbio.3002726

Yang, W., et al. (2024) Selection of experience for memory by hippocampal sharp wave ripples. Science 383(6690): 1478–1483. 10.1126/science.adk8261

Yiu, A.P., et al. (2014) Neurons are recruited to a memory trace based on relative neuronal excitability immediately before training. Neuron 83(3): 722–735. 10.1016/j.neuron.2014.07.017

Yvert, G. (2014) 'Particle genetics': Treating every cell as unique. Trends Genet. 30(2): 49–56. 10.1016/j.tig.2013.11.002

Zakharov, V.M., et al. (2020) Fluctuating asymmetry, developmental noise and developmental stability: Future prospects for the population developmental biology approach. Symmetry 12(8): 1376. 10.3390/sym12081376

Zaremba-Niedzwiedzka, K., et al. (2017) Asgard archaea illuminate the origin of eukaryotic cellular complexity. Nature 541(7637): 353–358. 10.1038/nature21031

Zecha, J., et al. (2023) Decrypting drug actions and protein modifications by dose- and time-resolved proteomics. Science 380(6640): 93–101. 10.1126/science.ade3925

Zhang, B., DeAngelis, D.L. (2020) An overview of agent-based models in plant biology and ecology. Ann. Bot. 126(4): 539–557. 10.1093/aob/mcaa043

Zhang, L., et al. (2020) FRET kinase sensor development reveals SnRK2/OST1 activation by ABA but not by MeJA and high CO_2 during stomatal closure. eLife 9(1): e56351. 10.7554/eLife.56351

Zhang, X., et al. (2025) Reconstitution of synaptic junctions orchestrated by teneurin-latrophilin complexes. Science 387(6731): 322–329. 10.1126/science.adq3586

Zhang, Z., et al. (2025) A panoramic view of cell population dynamics in mammalian aging. Science 387(6731): eadn3949. 10.1126/science.adn3949

Zhao, J., et al. (2024) Structural basis of odor sensing by insect heteromeric odorant receptors. Science 384(6703): 1460–1467. 10.1126/science.adn6384

Zhou, Y., et al. (2009) CREB regulates excitability and the allocation of memory to subsets of neurons in the amygdala. Nat. Neurosci. 12(11): 1438–1443. 10.1038/nn.2405

Zhu, K., et al. (2023) An intrinsically disordered region controlling condensation of a circadian clock component and rhythmic transcription in the liver. Mol. Cell 83(19): 3457–3469.e3457. 10.1016/j.molcel.2023.09.010

Zhu, S., et al. (2024) Demixing is a default process for biological condensates formed via phase separation. Science 384(6698): 920–928. 10.1126/science.adj7066

Zhu, X., et al. (2025) Measurements of molecular size and shape on a chip. Science 388(6747): eadt5827. 10.1126/science.adt5827

Zilberman, D., et al. (2012) Adoption versus adaptation, with emphasis on climate change. Annu. Rev. Res. Econ. 4(1): 27–53. 10.1146/annurev-resource-083110-115954

Index

For the benefit of digital users, indexed terms that span two pages (e.g., 52–53) may, on occasion, appear on only one of those pages.

Figures are indicated by an italic *f*.

D

E

F

K

L